Do-It-Yourself Sustainable Water Projects

About the Author

Paul Dempsey is a DIY mechanic, bike rider, and a former magazine editor. He has written some 30 technical books, most of them about internal-combustion engines.

Do-It-Yourself Sustainable Water Projects

Collect, Store, Purify, and Drill for Water

Paul Dempsey

Tony Selby Illustrator

New York Chicago San Francisco Lisbon London Madrid
Mexico City Milan New Delhi San Juan Seoul
Singapore Sydney Toronto

The McGraw-Hill Companies

Cataloging-in-Publication Data is on file with the Library of Congress.

1 2 3 4 5 6 7 8 9 0 QFR/QFR 1 9 8 7 6 5 4 3

ISBN 978-0-07-179422-0
MHID 0-07-179422-0

This book is printed on acid-free paper.

Sponsoring Editor
Judy Bass

Acquisitions Coordinator
Bridget Thoreson

Editing Supervisor
David Fogarty

Copy Editor
Joe Cavanagh
D&P Editorial Services

Production Supervisor
Pamela A. Pelton

Composition
D&P Editorial Services

Proofreader
Don Dimitry
D&P Editorial Services

Art Director, Cover
Jeff Weeks

Indexer
WordCo Indexing Services, Inc.

Project Managers
Nancy Dimitry,
Joanna Pomeranz
D&P Editorial Services

This book is dedicated to Larilla Templeton
who looks after all of us—
Mica, Erica, Araceli, Bebe Joel, Ariel, and Ramón.

Contents

Introduction

The drought of 2011 that extended from northern Mexico through Texas, Oklahoma and into the Dakotas brought the water crisis home for millions of Americans. But it was not a complete surprise: like a bad lab report or a visit from the IRS, it confirmed something we had long suspected.

Supplies of clean drinking water are becoming increasingly problematic. In 2002, 8% of the world's population labored under extreme water scarcity, as defined by drinking from sewage-laden ditches, walking for hours a day to communal wells, and similar deprivations. By mid-century, 40% of the world population, or some four billion people, are expected to be in the same predicament. Water will replace oil as the rationale for war. People can live without SUVs and jet travel, but no life—human or otherwise—can survive without water.

If climate models are correct, the ambient temperatures in the American Southwest will increase 5–7°F. Warm air holds more moisture, which means that the area will see less rain and the rain that does fall will quickly evaporate. Major river basins, including the Rio Grande, Colorado and Missouri will experience severe reductions—sometimes as great as 20%—in flow rates. In short, the Southwest will revert to its natural desert state.

According to the American Association for the Advancement of Science, groundwater, which supplies almost a third of the irrigation for agriculture, is being depleted 160% more rapidly than the aquifers recharge. The Ogallala, which contains fossil water left over from the last Ice Age, will be a memory during the lifetime of most readers. Short of some technological miracle, the Great Plains will return to grassland.

The outlook for municipal water supplies is not much better. The last upgrade of these systems occurred during the boom years following the Second World War. Treatment plants and distribution networks have long since exceeded their 50-year design life. While data is hard to come by, it appears that municipal systems lose, on the average, about 30% of the water they pump to leaks. In some cities the figure is 50%.

Older cities in New England and the Midwest have their storm drains cross-connected to sewage lines. Heavy rains—rains that climate change produces—overwhelm the treatment plants, and raw sewage enters the potable water mains.

Water-starved El Paso is one among several American cities that has turned to desalination. Because of the energy requirements, freshwater obtained in this manner costs an order of magnitude more than water obtained from aquifers and surface sources. Other cities are now recycling sewage, which, aesthetics aside, is also an expensive proposition.

Some idea of the desperation that water professionals feel can be had from the schemes they promote. Authorities in Southern California, Arizona, and Nevada have offered funds to build a 50-million gallon per day desalination plant in Playa de Rosaria. None of the plant's output would be exported to the U.S. Instead, the Mexican government would renounce some of its claims on Colorado River water.

The Southern Nevada Water Authority is seriously lobbying for a scheme to reduce demand on the Colorado River by recharging the Ogallala with flood waters from the Mississippi and Missouri Rivers. Should this project go through, it would be an engineering feat on par with the Aswan Dam.

Municipal systems must be updated and expanded with a cost estimated as high as one trillion dollars. Where the money will come from, no one knows. But family water bills will not be immune.

This book describes ways we can reduce or eliminate our dependence upon public water supplies. The most cost-effective way of doing this is simply to use less with low-flow toilets and washing machines, and ecologically compatible gardening. Other chapters describe how to drill for water and harvest rain and air-conditioner condensate. Pumps, the essential tool for working with water, receive intensive focus. Another chapter explores ways to harness solar, wind, and human power.

This is not a technological wonder book. I have tried to be objective, discussing the tradeoffs involved with solar cells and water-sparing appliances and the ways to avoid the frustration that sometimes arises when we rely upon outside contractors. Wherever practical, do-it-yourself (DIY) solutions are described and photographed. DIY projects include a hand-operated well pump capable of lifting water 100 feet or more, a composting toilet, and a rainwater harvesting system. Some of these projects are accompanied with detailed step-by-step instructions. A drawing or a few photos suffice for others, since how you proceed depends upon available materials and skills. However, special techniques, such as heat-forming PVC or where to find non-standard parts, are described in detail.

Water is a big subject with thousands of ramifications. No single book or single author can do justice to it. References in the text, most of them keyed to cost-free sources, will add immeasurably to your understanding of the magical substance that has gifted us with life.

Paul Dempsey
Boca del Rio, Mexico

1

Glimmers of Light

Water is strange, almost magical stuff for which there is no alternative.

Because water is such an excellent solvent, living cells depend upon it for oxygen, essential electrolytes, and nutrients. It also dissolves and carries off carbon dioxide and other waste products generated as cells convert sugars, fats, and glucose into energy.

Water absorbs heat better than any known substance except ammonia. That ability plus water's very high latent heat of evaporation makes for a very efficient cooling system. Each gram of evaporated sweat carries off 600 calories of heat. These characteristics are the result of the strong bonds hydrogen exerts between adjacent oxygen molecules. The same mechanism explains the high surface tension water exhibits and the ability of this marvelous liquid to overcome the force of gravity in narrow tubes. Trees and other vascular plants depend upon this capillary action for survival. Unlike most substances, water expands when cooled—ice has about 92 percent of the density of liquid water; consequently, ice floats, insulating the water below and enabling aquatic life to survive without freezing.

A really alien intelligence might look at human beings and other earthly organisms as ambulatory water containers. Water accounts for about two-thirds of the mass of our bodies. A 5 or 6 percent loss of water induces grogginess and headaches.

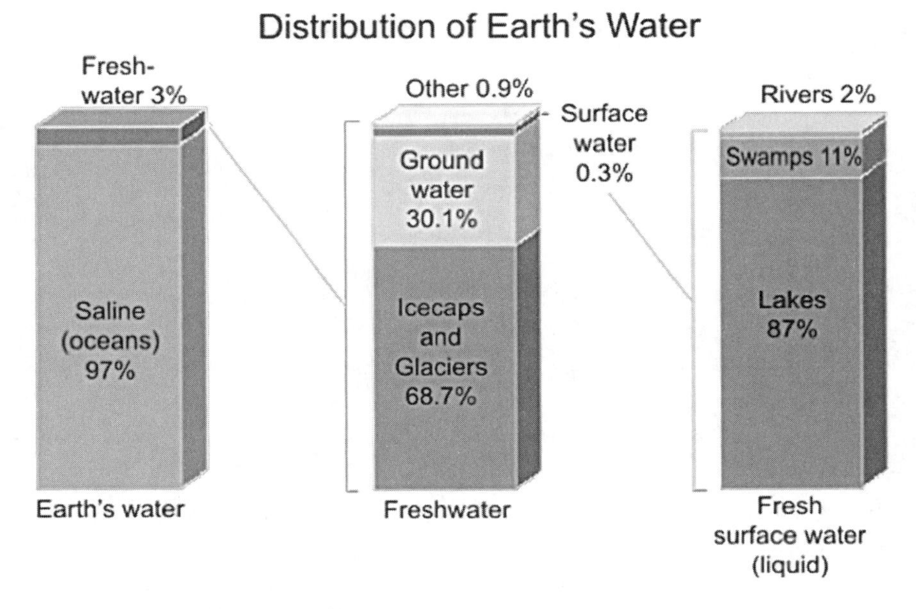

Figure 1-1. *How the Earth's water supply is allocated.* USGS

The urine turn dark yellow. At between 10 and 15 percent, we are in serious trouble—muscles become spastic, the skin loses elasticity, and vision dims. Death awaits.

The U.S. Geological Survey estimates the water content of our planet at 1,368,000,000 cubic kilometers (km^3) (Fig. 1-1). The oceans that cover almost three-quarters of the Earth's surface contain 97 percent of the water, which barring energy-intensive desalination, is almost useless for land-based life forms. The freshwater we need accounts for only 3 percent of the total, most of it locked in glaciers and permanent ice caps. For accessible freshwater, we must tap underground aquifers or drain rivers, lakes, and swamps. Surprisingly, the richest store of available water lies under our feet. Drill deep enough almost anywhere, and you will strike freshwater.

The sun provides the energy to evaporate water from the ocean, rivers, and soil, which then collects as vapor in the atmosphere (Fig. 1-2). As the warm vapor rises, it cools and condenses into clouds that further condense into rain, hail, and snow. Most precipitation falls on the oceans where it is immediately available for evaporation. A small percentage goes into semipermanent storage as glaciers and mountain ice caps.

Precipitation that falls on land either infiltrates into the ground or else skates over the surface as runoff. Most runoff returns to the ocean via rivers and streams. Infiltrated water tends to remain near the surface of the soil, where it awaits evaporation. Some tiny fraction of precipitation finds its way into aquifers, consisting of saturated subsurface rocks. Except for freshwater springs, aquifer water remains in place until tapped by humans.

If we disregard the almost infinitesimal amount of water than escapes the atmosphere, the amount of water on the planet does not change. However, there is a timescale involved. Most water use is consumptive, meaning that it does not immediately return to streams and rivers as runoff. If a

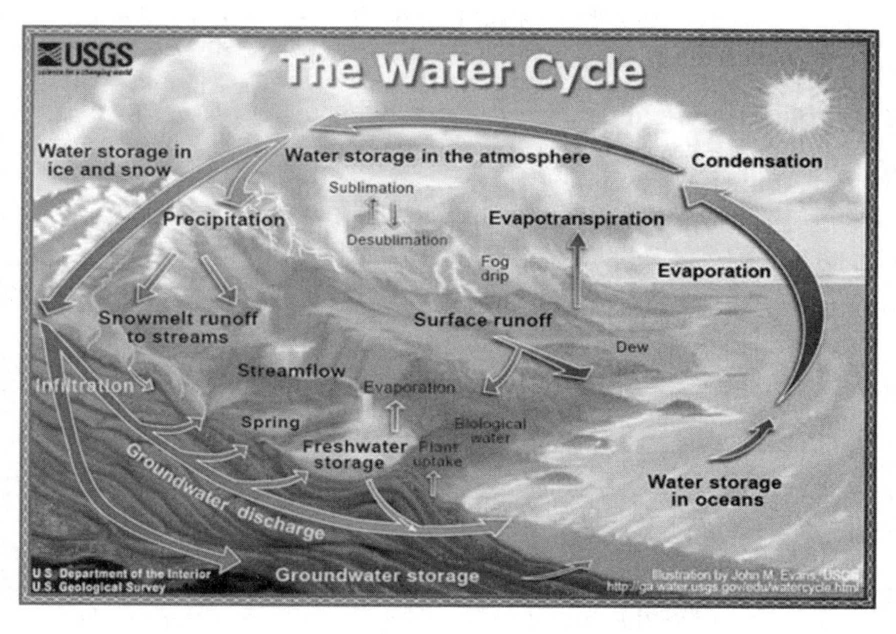

Figure 1-2. *The water cycle, driven by heat from the sun, endlessly repeats itself.* USGS

farmer exercises care, the water he uses for irrigation remains on his fields where it evaporates or transpires from plants. Eventually the closed water cycle will again make this water available, but the process is slow. Replenishing aquifers can take millennia.

Inequalities in distribution, coupled with excessive draws on groundwater reserves and contamination by agricultural and industrial wastes add up to the most severe water crisis in our history.

Worldwide Water Use

The seven billion people that currently inhabit our planet appropriate more than half of all accessible freshwater contained in rivers, lakes, and aquifers. About three-quarters of this water goes for irrigation; industry absorbs a fifth; and the 8 percent or so that remains is used for toilets, showers, washing clothes, and other domestic purposes.

One often hears that population drives water use. That is true only in a limited sense. Among impoverished populations, water consumption tends to be a zero-sum game: each additional mouth subtracts from the water available. Richer countries have more water to start with and so have room to maneuver. While data are incomplete, water demand in the twentieth century appears to have increased twice as fast as population growth. This suggests that affluence, or more exactly, habits associated with affluence, can drive water consumption even faster than population growth. The United States leads the world in per capita water consumption. Citizens in other developed countries consume less than half as much freshwater. Reducing water consumption does not necessarily come at the cost of economic development.

According to the United Nations, a per capita consumption of 1700 cubic meters (m^3) per year is necessary to establish conditions for a decent human life. Most of this water goes to feed the people involved: rice has a water cost of as much as 5000 liters per kilo; sugar, maize, and wheat are not far behind. Meat cattle are off the scale. A McDonald's hamburger has a water cost of 11,000 liters. Biofuels impose a cost of 1000 times their volume.

What the UN calls "water scarcity" occurs when consumption falls to less than 1000 m^3 per capita per day. At 500 m^3 or less, water scarcity becomes "absolute." Millions of people in sub-Saharan Africa, the slums of Asiatic cities, and the Middle East drink from roadside ditches, animal troughs, or sewage-polluted streams. In Africa and Latin America, women, many of them girls of school age, walk for hours every day to fetch water. The girls grow up illiterate and vulnerable. As Yeni Bazan, a 10-year-old girl in El Alto, Bolivia, said, "Of course, I wish I were in school. I want to learn to read and write.... But how can I? My mother needs me to get water."

By all accounts, the impacts of population growth, climate change, and industrial activity will make things even harder for people in developing countries, with absolute water scarcity the norm for 85 percent of the population in sub-Saharan Africa and tens of millions more in South Asia, the Middle East, and Latin America.

Unlike oil, freshwater is not fungible. No one is willing to pay for the tankers and pipelines that could, to some small

Metrics

Water professionals almost universally use the metric system.

Length

1 meter = 3.28 feet
1 foot = 0.305 meter
To convert meters to feet, multiply by 3.28. Thus
50 m = 50 × 3.28 = 164 ft.
To convert feet to meters, multiply feet by 0.305. Thus
100 ft = 100 × 0.305 = 30.5 m

Volume

1 liter = 0.264 U.S. gallon
1 gallon = 3.78 liters
To convert liters to U.S. gallons, multiply the number of liters by 0.264. To work the conversion the other way, multiply the number of gallons by 3.78 to obtain liters.
1 cubic foot = 7.48 U.S. gallons or 28.32 liters.
1 cubic meter = 1000 liters or 264.2 gallons.
1 cubic kilometer = 1,000,000,000,000 liters or 264,172,052,358 gallons

degree, equalize distribution, although every few years a scheme is broached to tow icebergs to the Red Sea. But we can support nongovernmental organizations (NGOs) in their work to bring water to the needy. You can find contact information for the most active of these organizations at the back of the book.

Here at Home

The water crisis also affects North America. Nineteen climate models independently predict that the drought extending from northern Mexico into the southwestern United States will continue into the next century. According to Richard Seager of Columbia University's Lamont-Doherty Earth Observatory, "the levels of aridity of the recent multiyear drought, [similar to] the Dust Bowl and 1950s droughts, will, within coming years or decades, become the new climatology of the American Southwest."[1] Dr. Seager added that he did not mean that the dust storms of the 1930s, which were triggered by poor agricultural practices, would be replicated. But the economic effects, with thousands of farmers and ranchers displaced, will be the same. Data provided by the U.S. Weather Service gives a more detailed picture (Fig. 1-3).

Underground reservoirs, such as the Ogallala aquifer that extends from West Texas to South Dakota, hold 90 percent of the world's accessible freshwater. The equivalent of Lake Huron, the Ogallala makes wheat growing, ranching, and large-scale urban settlement possible.

Extraction began in the 1930s and accelerated rapidly as more efficient deep-well pumps were developed. Since the 1940s water levels in the aquifer have dropped 100 ft. The American Association for the Advancement of Science (AAAS), reports that the Ogallala "will likely become nonproductive within the next 40 years." Rain and snowmelt will someday restore the aquifer, but it will be a long wait, since recharge averages an inch a year.

Nor is the problem limited to the Ogallala. The AAAS found that "groundwater, that provides 31 percent of the water used in agriculture, is being depleted up to 160 percent faster than its recharge rate." This is called living off one's capital.

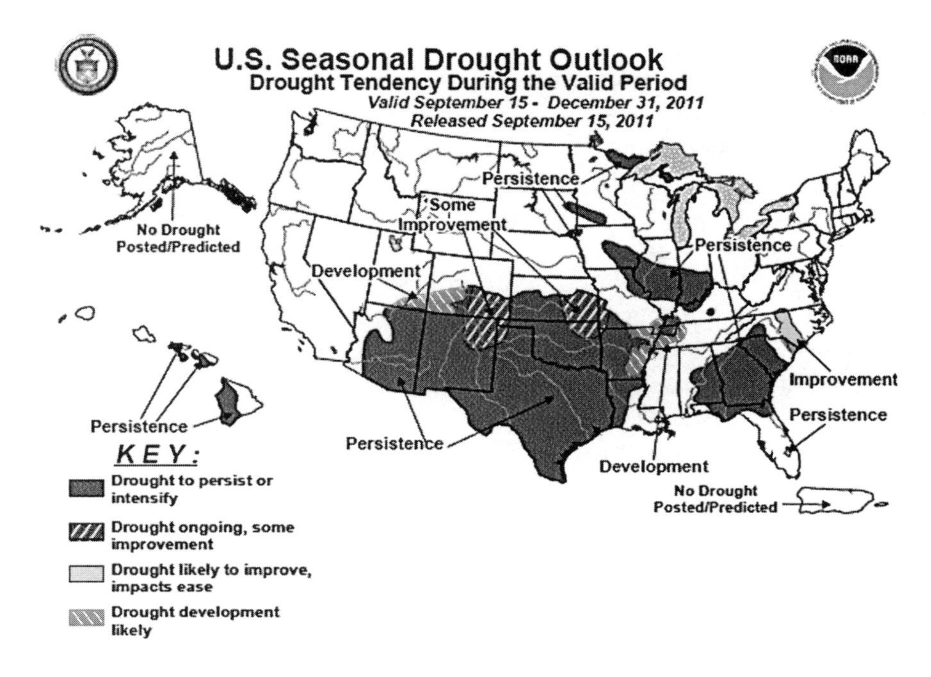

Figure 1-3. *Drought appears to be the new norm for the American Southwest and other regions of the country. The map above depicts large-scale trends based on subjectively derived probabilities guided by short- and long-range statistical and dynamic forecasts. Short-term events—such as individual storms—cannot be accurately forecast more than a few days in advance. Use caution for applications, such as crops, that can be affected by such events. "Ongoing" drought areas are approximated from the Drought Monitor (D1 to D4 intensity). For weekly drought updates, see the latest U.S. Drought Monitor. NOTE: The gray improvement areas imply at least a 1-category improvement in the Drought Monitor intensity levels, but do not necessarily imply drought elimination.* U.S. Weather Service

Contamination

Not very long ago the United States led the world in the quest for safe drinking water. Cleanup began in 1866 when the New York Board of Health was tasked with eliminating cholera and other waterborne epidemics that periodically swept through the city. The effort was successful and other American municipalities followed suit, building water-purification and sewage-treatment plants. Chicago stopped using

Lake Michigan, the source of the city's drinking water, as a sewage dump; Cleveland extended its water-intake lines four miles out into Lake Erie. In 1880, only 1 percent of the American population had access to filtered water. Sixty years later the figure had increased to 50 percent.

In 1972 Congress passed the Clean Water Act to regulate more than 100 pollutants discharged by refineries, chemical plants, and other manufacturing operations. The companion Safe Drinking Water Act limits or prohibits 91 chemicals and pathogens in potable water.

But the consensus that made clean water a national goal appears to have evaporated. In the 5-year period ending in 2009, 20 percent of U.S. water-treatment plants violated provisions of the Safe Drinking Water Act. Some 49 million Americans drank water laced with illegal levels of arsenic, radioactive uranium, tetrachloroethylene (a dry-cleaning solvent), and other carcinogens. During the same period, 205 New York State treatment plants delivered water containing sewage bacteria. The Clean Water Act has also been honored more in the breach than the observance. The Environmental Protection Agency reports that a half-million illegal chemical discharges occurred during the period under discussion.[2] Almost none of the violations were prosecuted.

Even so, the EPA believes that nonpoint pollution—pollution with a generalized, difficult-to-pinpoint source—represents an even greater threat to the water supply. Runoff from rain and snowmelt absorbs fertilizer, animal droppings, pesticides, oil, and industrial wastes and deposits these contaminants into lakes and rivers. Nonpoint pollution made 40 percent of the streams and rivers the agency surveyed unsuitable for swimming or fishing. This sort of pollution eludes effective control.

Climate Change

As the Earth warms,

- Dry areas of the world, such as the American Southwest, will become dryer and hotter. The Bureau of Reclamation projects a 5- to 7-degree Fahrenheit (F) temperature rise

for western U.S. states during this century. The combination of higher temperatures and retreating snow packs will reduce the flows in major river basins by as much as 20 percent. What will happen to the Rio Grande and Colorado hardly bears thinking about. At the same time, regions in the high latitudes and in the tropics will experience intense precipitation of the kind that causes rapid runoffs, soil erosion, and flooding.
- If the models are correct, sea levels will rise as much as 1.6 meters (m) (5¼ feet [ft]) by the end of the century and peak at 6 m (almost 20 ft) as the Greenland and Antarctic ice caps melt. A worrisome sign is that the northwest and northeast passages—the Arctic Ocean shortcuts to Asia sought by the sixteenth century explorers Frobisher and Willoughby— have opened. The summer ice has thinned enough to allow commercial traffic along these routes.

In view of the foregoing, it's difficult to escape the conclusion that water scarcity will define this century in the same way that oil defined the twentieth century and coal the century before. Men will search for it, lay claim it and fight over it.

Some measure of the seriousness of the situation can be had from the way that water agencies in Southern California, Las Vegas, and Phoenix are pushing for construction of a massive 50-million gallon-per-day desalination plant in Playa de Rosarito, Mexico. Plans call for the Americans to help pay for the plant, and in return, Mexico would surrender some of its claims to Colorado River water. Critics see the proposal as outsourcing—a way to avoid environmental regulations.

Pat Mulroy, the general manager of the Southern Nevada Water Authority and the Las Vegas Water District and one of the most powerful women in the country, has agitated for years for the construction of a massive flood-control project for the Mississippi and the Missouri rivers. The surplus water would recharge the Ogallala aquifer, reducing irrigation demands on the Colorado River, upon which Las Vegas depends. What this project would do to the already disappearing wetlands on coastal Louisiana is a matter of conjecture.

But enough of the bad news. There is reason for hope.

The Way Forward

Large scale problems do not require large-scale solutions—they require small-scale solutions within a large-scale framework.

—David Fleming

That quote from one of the founders of the British Green Party may strike one as pixie dust, as the kind of facile optimism associated with street-corner preachers and health cranks. But consider. One poor fruit seller in Tunisia set the fire that destroyed a half-dozen Middle Eastern dictatorships; a barely literate Michigan farm boy put America on wheels; a college dropout made possible the operating system used to write this book. Individuals, working alone or within small like-minded groups, are the catalysts of change. Governments and corporations arrive late on the scene to scale up and institutionalize change.

The water crisis can be attacked from a thousand directions—God knows there's plenty to do. But the best place to begin is with a few simple steps to reduce your own consumption:

- Fix leaking faucets
- Replace thirsty fixtures with low-flow types
- Wash full loads of clothes
- Do not over irrigate lawns and garden plants.

The results will be immediately apparent in the monthly water bill. To achieve greater savings:

- Replace old-model washing machines and toilets with low-flow models.
- Redo the yard and garden on the Xeriscape model (see Chapter 2, page 21) with plants and trees adapted to local conditions.
- Fabricate or purchase catchment systems for rainwater and air-conditioner condensate.
- Readers in rural areas can drill their own wells and power the pump with solar or wind energy.

Personal involvement in the work of collecting, drilling for, purifying, and storing water seems essential. As Giambattista Vico (1668–1744) put it, "The criterion and rule of the true is to have made it." In other words, we can truly understand only what we fabricate. If you want to learn about water, you have to work with it, using whatever materials, skills, and imagination are at hand. This experience gives one a kind of authority that is instinctively recognized. Neighbors and friends will applaud your efforts and may imitate them.

The next step is to become active politically. Politicians react when a bridge falls down or an athletic team wants a new stadium, but few care about municipal water treatment plants or distribution networks. The 880,000 miles of mostly buried pipe that supplies North Americans with drinking water spring a quarter-million leaks every year, which is not surprising. Some of the pipe is more than a hundred years old. On the average, U.S. and Canadian municipal water systems lose about 30 percent of the water they pump. In some cities, the figure is 50 percent.

We are also witnessing a revolution in water treatment. San Diego has recently opened a facility that converts a million gallons a day of blackwater—raw sewage—into potable water. Drinking 100 percent reclaimed sewage is a new development, impelled by drought and population pressures.

The technology is also used on the International Space Station, as well as in Singapore, where rain is the only source of freshwater. The San Diego plant combines chemical disinfection, using hydrogen peroxide and infrared light, with reverse osmosis to kill pathogens and remove solids. Other cities, including Miami and Denver, are considering doing something similar.

However, multiple filtration, while less expensive than desalination, costs considerably more than tapping groundwater. We must overcome our instinctive abhorrence to toilet-to-tap water. People like to think that their drinking water comes from crystal-clear mountain streams.

According to a recent report by the National Academy of Sciences, sewage recycling would enable coastal cities to increase their domestic freshwater supplies by 27 percent. Yet only 0.1 percent of potable water in the United States originates solely from sewage. Of course, cities that discharge treated sewage into waterways merely pass the diluted effluent along to downstream consumers. But that somehow is considered more acceptable that drinking 100 percent sewage-sourced water. As one old gentleman, quoted in *The New York Times,* said, he wouldn't let his cat drink toilet-to-tap water.

Those of us who are aware of the dimensions of the crisis need to speak out on these issues in order to make our voices heard in the community and, ultimately, at the national level. We have to make certain that the old gentleman's cats do not have the final say.

Political action requires skill and organization. One of the savviest organizations is the Waterkeeper Alliance that grew out of efforts by New York fishermen to protect the Hudson River. Today the Alliance has some 190 worldwide chapters, each dedicated to guarding a local waterway. Thanks to the efforts of these men and women, fisheries have been restored, millions of dollars of compensation money has been collected, and hundreds of polluters have been brought to trial.

Interested readers should contact:

Waterkeeper Alliance
17 Battery Place Suite 1329
New York, NY 10004
United States of America
Tel. 212-747-0622
Fax 212-747-0611
info1@waterkeeper.org
www.waterkeeper.org

Notes

1. Associated Press, April 5, 2007.
2. The New York Times series "Toxic Waters" explored the health risks associated with U.S. drinking water in exhaustive detail. These articles can be found at http://www.nytimes.com/2009/12/17/us/17water.html?pagewanted=all.

2

Using Less

NAEUS

The North American End Use Study (NAEUS) is the best account we have of the way households use water. University of Colorado students metered consumption at 10-second intervals for 100 middle-class homes in 12 cities across the country. In all, the students collected 28,000 days of data.

Consumption per household averaged nearly 147,700 gallons per year (gpy), or 69.3 gallons per capita per day. Active adults need to drink about a half-gallon of water a day. The other 68.8 gallons represent opportunities for conservation.

Lawn and Garden

The population shift to warmer and drier parts of the country, together with the larger, more intensively cultivated lawns favored by suburbanites, has increased outdoor water use. The long growing season in the South and the popularity of automatic sprinkler systems contribute to the problem.

Lawns and gardens account for about 60 percent of the water the average household uses. In places like Tempe, Arizona, the figure is 80 percent. Grass clippings, the harvest from nearly 50,000 square miles of lawn, are our largest irrigated crop. Corn comes in a distant second.

Las Vegas, which receives four inches of rain in a good year, or a tenth as much as Chicago, has become a laboratory for outdoor water conservation. The Las Vegas Valley Water District decorates roadsides with boulders, compensates farmers to take land out of cultivation, and pays homeowners $1.50 a square foot to replace their lawns with gravel. As a result, the city's water consumption has dropped by nearly a third, during a period when its population grew by nearly 400,000.

There are, however, less draconian ways to conserve.

- NAEUS found that homeowners over-irrigate, using on the average of four times more water than their plants need. Unless you live in the high desert, turf and ornamental plants require no more than a weekly watering. Too much water discourages the deep root growth that healthy plants require. Try to irrigate evenly and minimize runoff.
- Plants absorb water at root level, which for grass is six or eight inches below the surface. Take a soil sample from that depth. If you can roll the soil into a loose ball, the grass has sufficient moisture. If, on the other hand, walking on the lawn leaves visible footprints, the grass and, one can assume, other yard plants need watering.
- Irrigate in the late evening or in the early morning hours. Probably the greatest advantage of a permanently installed irrigation system is that the chore can be done at night.
- Rotary spray nozzles work well for lawns. Newer designs throw large droplets in low trajectories that inhibit evaporation. Plants and trees respond to soaker, or drip, hoses that deliver water no faster than it infiltrates the soil. Microspray nozzles fitted to permanent irrigation systems have a similar effect.
- Shelter plants and trees with compost or mulch. Strangely enough, compost costs less than mulch, although it takes

much longer to process. Berms and steps cut into hillsides reduce runoff and encourage infiltration.

- Set the mower blade high so that turf can create a protective covering.
- Give priority to the trees during dry spells. Grass may go dormant and turn brown, but it can always be reseeded. Trees require decades to reach maturity.
- Go easy on the chemicals. Lawns are repositories of something like a quarter of all herbicides sold in the United States. We track these toxic chemicals into our houses on our shoes and some fraction of the runoff finds its way into our drinking water. The same holds for the nitrates in fertilizers, which are major ground and surface water pollutants.

Xeriscape

In response to a 1981 drought, the Denver Water Department developed a set of gardening techniques known as Xeriscaping. The term combines the Greek "xeros," or dry, with "scape," an old English word that survives in "landscape" and "seascape."

Xeriscaping replaces traditional turf with native plants, adapted to the climate and rainfall variations. Once put into place, these plants look after themselves and very little maintenance is needed. The wonderful tulips pictured in Fig. 2-1 were grown under semi-desert conditions in the Denver Botanic Gardens, without herbicides or artificial fertilizers.

California has made Xeriscaping or some approximation of it mandatory for new construction in water-stricken areas. State officials point out that these techniques also save energy. As things currently stand, water treatment and pumping facilities account for about 20 percent of California's electrical consumption.

Xeriscaping includes these elements:

- **Xeric plants**—a variety of vines, shrubs, ground covers, and perennials adapted to low moisture conditions form the basis of the garden. Many of these plants have long

Figure 2-1. *The Denver Botanic Gardens, which attracts tens of thousands of visitors a year, is a Xeriscape demonstration project.*
Scott Dressel-Martin

blooming seasons and leaves that undergo a magical transformation in the autumn.

- **Turf**—grassy zones should be confined to areas near the house and consolidated for easy mowing. Play areas for children can be covered with sand or mulch, as can paths for garden access.
- **Layout**—plants are grouped according to their water needs. Areas frequented by humans, such as entranceways and patios, receive the most water.
- **Soil conditioning**—tilling is essential to encourage root growth and hold moisture.
- **Mulch**—a variety of organic and non-organic mulches are used to conserve water, lower soil temperatures, and inhibit weeds. A third of the soil volume should consist of organic mulch, with compost blended in to a depth of six or more inches.

- **Irrigation**—little and infrequent watering will encourage root development.

Indoors

Almost half of water consumption takes place indoors.

Leaks

Dripping faucets, leaking toilets, and porous piping can use up to 10,000 gallons a year, or enough water to fill a moderately sized swimming pool. A drop a second adds up to 3000 gallons a year.

New washers usually put a stop to faucet drips; "pull-down" faucets, the kind that control hot and cold water with a single lever, require a new cartridge, available for name-brand units from plumbing supply houses. To make toilet-tank leaks visible, put a little food coloring in the tank. If, after a few minutes, water in the bowl takes on color, replace the flapper valve.

Once the obvious leaks have been repaired, turn off all water in the house, including the refrigerator's ice maker. Write down the meter reading, and check the reading again after 30 minutes or so.

Line Pressure Reduction

Municipal distribution systems work by gravity. If you live near a storage tank, your line pressure can reach 100 pounds per square inch (psi). On the far edge of the network, pressure drops to as little as 20 psi.

If you have normal pressure—60 psi or higher—adding a pressure-regulating valve at the service connection may be appropriate. These adjustable valves reduce the flow by a third, prolong the life of plumbing, and negate the need to purchase low-flow faucets and showerheads. However, washing machines and toilets use a preset amount of water, regardless of line pressure.

Most regulating valves incorporate a diaphragm-operated disc (Fig. 2-2). High water-main pressures move the disc closer

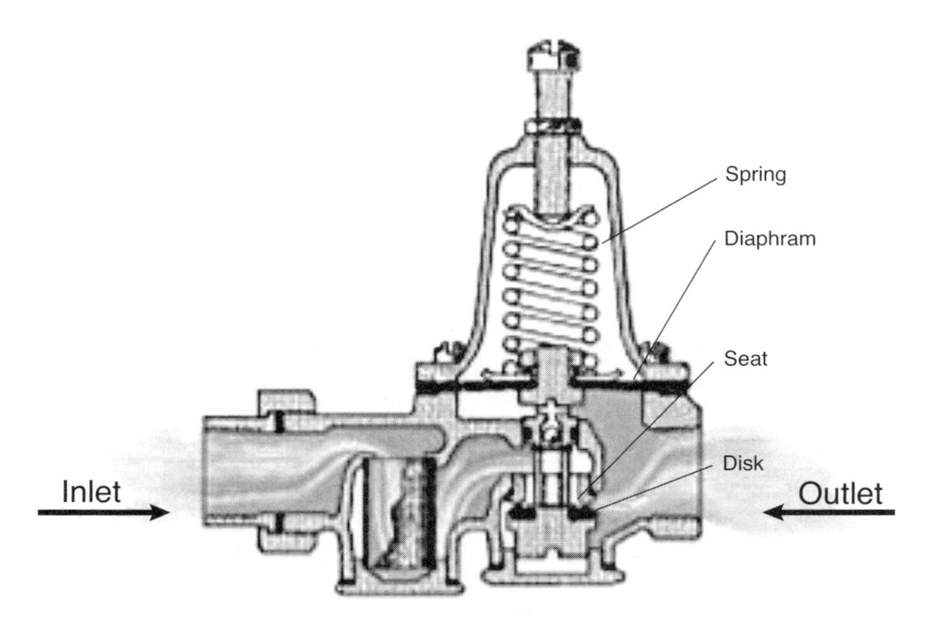

Figure 2-2. *Diaphragm-type pressure regulating valve.*
Watts Water Technologies

to its seat to restrict flow and pressure. Should pressure drop, the valve opens to permit more flow. Elegant and simple.

Sizing the valve requires a bit of thought, since the pressure drop varies with the flow rate. Figure 2-3 shows how Watts valves are sized. Other manufacturers have their own curves and may use a different terminology, but the principle is the same.

A single valve, installed directly behind the meter and in series with it, is sufficient for residences and small businesses. Apartment complexes, where water flow can vary from almost nothing to 600 gpm (gallons per minute) in early morning hours, benefit from two regulating valves, plumbed in parallel (Fig. 2-4). This configuration provides more flow than possible from a single valve and permits either valve to be serviced without shutting down the system.

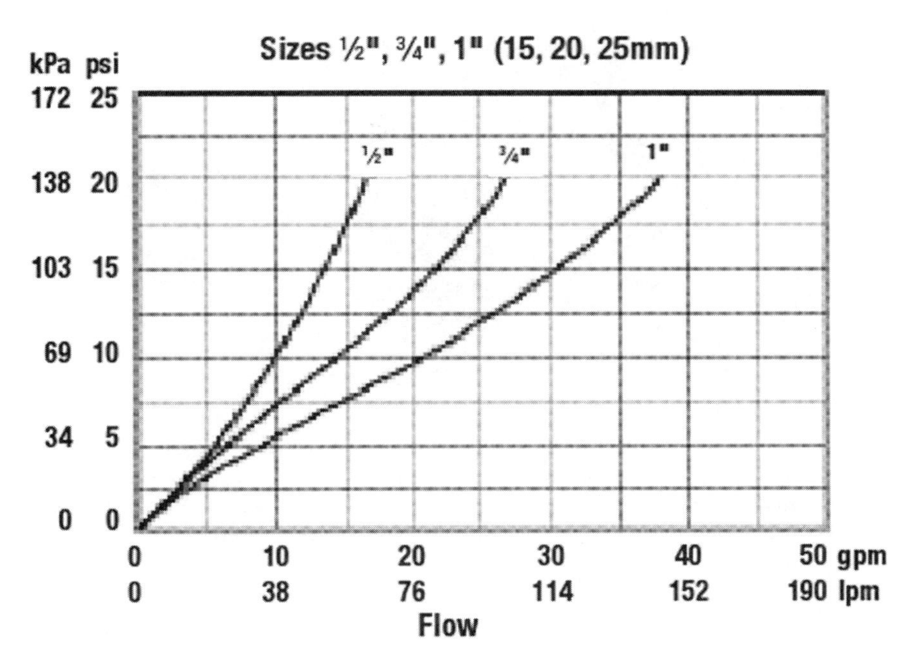

Figure 2-3. *A regulator valve imposes an internal pressure drop that subtracts from its initial pressure setting. For example, if the initial setting is 50 psi, a ¾-in. regulator will lose about 13 psi when it flows at 20 gpm. A larger, 1-in. regulator drops only 10 psi at the same flow rate.* Watts Water Technologies

Low-Flow Showerheads and Faucets

Showerheads manufactured before 1992 flowed more than five gallons a minute. The current standard is 2.2 gpm. Some showerheads produce less flow, although it appears that 2.0 gpm is the comfort limit.

Place a one-gallon bucket under the shower and open the spigot wide. If the bucket fills in 20 seconds or less, replace the head with a low-flow type. These heads come in two styles: aerating heads break up the water stream with air bubbles; laminar heads generate a cone of tiny, high-velocity streams. Either type costs between $15 and $30.

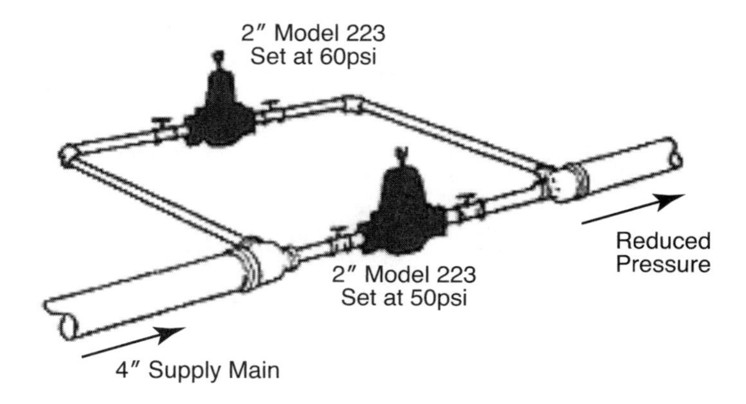

Figure 2-4. *A parallel hookup provides more stable pressure and redundancy. Note that the lower set valve is recommended to be located on the main run with the higher set valve located on the saddle for easier maintenance of this 100 percent used valve.*
Watts Water Technologies

Water Sense lavatory faucets deliver 1.5 gpm at 60 psi to save the average household 500 gallons per year (gpy). Standard kitchen faucets flow 2.0 gpm. A $5 aerator threaded on the spout throttles the flow down to 1.5 gpm.

Inline Water Heaters

An inline heater installed near the shower outlet provides hot water almost instantly and eliminates the small but real risk associated with hot water tanks. But inline heaters are not inexpensive and installation can be quite a project. A solar heater may be the better option (Fig. 2-5).

Energy Star Washers

Conventional top-loading washing machines average about 41 gallons per load. Washers that the EPA certifies as meeting Energy Star norms reduce water consumption by a third or more and save on the use of electricity and detergent in the bargain. Most of these washers are front loaders, although several recent models load from the top.

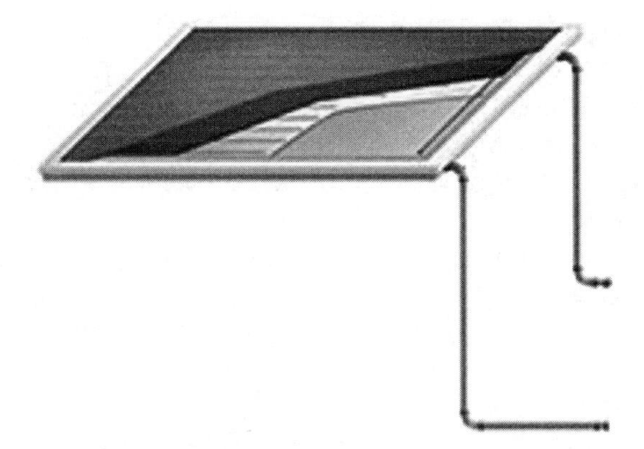

Figure 2-5. *A solar flat-plate collector is one step up from coiling a water hose in the sun. The collector consists of copper tubing inside a box covered with tempered glass. A hundred-foot coil of standard ¾-in. Type K tubing contains 2.7 gallons, or enough for a quick shower. A hot-water heater tank and pump will provide more storage. Install a cutoff valve and a drain to prevent freezing.*
Energy Star, U.S. government

The paperwork accompanying these machines includes something called the water factor (wf). If the washer requires 24 gallons per cycle and has a drum volume of 3.0 cubic feet (cu. ft.), then the wf is 24 ÷ 3.0 = 8.0. The lower the number, the less water used. As of 2011, Energy Star washers have a wf of 6.0 or less.

Conventional top-loading washers have a vertical drum that fills and empties several times during the wash cycle. Most Energy Star washers use a horizontal drum accessed through a door on the front of the machine. The drum contains only a few inches of water. As the machine runs, the drum reciprocates, lifting and dropping the clothes into the water below. During the spin cycle the drum rotates at speeds in excess of 1000 rpm. Clothes come out almost dry.

Energy Star washers and their European AAA counterparts have not been an unqualified success. Customers complain about leaky door seals, $500 circuit boards, severe vibration and bearings that, on some models, are not individually replaceable. Should the bearings go, the customer has to buy an entire drum assembly. According to the U.S. National Society of Homebuilders, washing machines have a ten-year life span. A British government study puts it at seven years. Assuming a random distribution, half of the machines fall short of the average and half exceed it.

In times within memory washing machines and other home appliances had life spans measured in decades. Hotpoint washers survive from the 1970s, and the worst that goes wrong with antique Maytags is that the wringer rollers harden with age.

The decline in customer satisfaction can be explained, in part, by the demands put on engineers to radically reduce water and energy consumption. Major redesigns hardly ever go smoothly. And these changes had to be made in the context of severe price competition. In 1947, a Bendix Deluxe retailed for $250. The same machine would cost $2577 in 2012 dollars. But Energy Star front-loaders, machines an order of magnitude more complex than the Bendix, sell for less than a $1000. Something had to give and that something was build quality. Economists call the phenomenon *stealth inflation.*

To put things into context, it is useful to remember that washers and other home appliances represent mature technologies. There are no breakthroughs, no magic bullets like those that have doubled computer processing speeds every six months. Instead of Silicon Valley, washers come out of border-town maquiladoras, Oriental sweat shops, and a few antiseptically clean plants in Sweden.

Readers in the UK might look at the ISE W1607, designed by a group of environmentally conscious engineers in Birmingham and manufactured in Sweden. The machine carries an A+++AA efficiency rating and comes with a 10-year parts and labor warranty, which is the best in the industry. Each component has been tested for the equivalent of 28 years of normal operation, although the factory expects that the average life will be 22 years. The perennial door-seal problem, the bane of front-loaders, appears to have been solved. And the drum bearings were originally designed to run for a million kilometers in Volvo trucks.

What is even more remarkable is that the company sells replacement parts at cost and makes service information available to independent repair shops. The ISE is a kind of open-source washing machine.

Potential buyers on this side of the Atlantic can begin their search with product reviews posted by Home Depot, Best Buy, Lowes, and Amazon. J.D. Power provides some indirect information about reliability, but neglects to include model numbers. Consumer Reports has some good general advice—avoid bling, purchase upper-mid-range models, and forego service contracts—but their reviews could focus more on long-term usage.

The Appliance Blog http://www.applianceblog.com/main forums/forum13/0 is a valuable resource for those of us who make our own repairs. This outfit also sells parts and posts links (buried in the "Comments") for repair procedures and manuals. Shop manuals for more popular models can be found at http://www.applianceaid.com/appliance-repair-manuals.php. You might also want to check out the parts catalogued by the Repair Clinic (http://www.repairclinic.com).

Flush Toilets

Over the course of a normal lifetime, nature calls us to the bathroom 140,000 times, or so someone has calculated. It's not surprising that NAEUS found that toilets consume more than 18 gallons of water per person per day. Washing machines come in a distant second.

In terms of gallons per flush (gpf), four types of toilets are found in the United States:

1. **Unregulated**—3.4 gpf, with some using as much as 6 gpf. If the toilet has been designed at all well, you can reduce consumption a bit by placing a brick or a bottle of water in the tank.
2. **Low-flush**—1.6 gpf. In 1994, the EPA mandated that all new commodes meet this standard. First-generation toilets were a plumber's nightmare, clogging sewer lines and requiring multiple flushes. Newer models work better.

3. **Water Sense**—1.28 gpf. Toilets meeting this standard are mandatory for U.S. government installations and are available to the general public.
4. **Ultra-low flush**—less than 1.28 gpf. Some manufacturers offer toilets that exceed the Water Sense standard. It appears that 0.6 gpf is the practical limit.

No-frill Water Sense toilets retail for around $300 and, according to the EPA, should pay for themselves in three years. Rebates, on the order of $50 or so, are available in many areas of the country. Ultra-low flush toilets generally start at around $500.

DIYers can reduce their water consumption by plumbing rainwater into the toilet holding tank, as described in the following chapter. It's also possible to redirect greywater from the shower, lavatory, and washing machine into the bowl and *not the toilet tank*. Greywater is both dirty and corrosive. However, the effort seems a bit quixotic: unless the washing machine runs almost constantly, grey flush water will have to be augmented with tap water. Asking the family not to flush so frequently can have a similar effect. You can remind them that not too many years ago, chamber pots were found in every bedroom.

Composting Toilets

Composting involves a host of biological and chemical processes that convert animal and vegetable matter into humus. It is nature's way of housekeeping, of dealing with the waste that all life forms generate. We merely assist nature by choosing a warm site, providing vents for oxygen, and by adding grass clippings and leaves or sawdust to maintain the nitrogen-carbon ratio. In warm climates, the process completes itself within a few months to produce a dry, fluffy loam that may, or may not, be safe to use on crop and pasture land (Fig. 2-6).

National governments and NGOs in poorer parts of the world consider composting toilets (CTs) a public health measure. Some 2.6 billion people, or nearly half the population

Figure 2-6. *Compost that has "fallowed" for 17 months in a large CT constructed by volunteers for Baltimore Jonah House. Except for shreds of toilet paper, the humanure came out perfectly. Note that the worker wears gloves and should be wearing the mask that hangs loosely around his neck.*

in these countries, live under conditions of extreme water scarcity. As the United Nations 2006 Development Program put it, without access to CTs people "defecate in ditches, plastic bags or on roadsides... and they collect water from drains, ditches or streams that might be infected with pathogens and bacteria that can cause severe illness and death." Figure 2-7 illustrates the miserable way sewage is handled at an elementary school in Afghanistan.

Authorities in the U.S. are less sanguine about composting toilets. Approval of CTs is only granted in the absence of sewer lines. Some state agencies go further and require that composting toilets conform to NSF/ANSI Standard 41, "Non-liquid Saturated Treatment Systems," a process that requires

Figure 2-7. *Sewage disposal for a newly constructed elementary school, near Kabul. Waste flows from latrines through an open ditch into the gully in the foreground.* USAID

six months of independently verified tests. CT owners in Oregon can be asked to post a $5000 bond.

Even so, the industry is booming (Fig. 2-8), and tens, or even hundreds of thousands, of Americans have constructed their own versions of these toilets.

Figure 2-8. *One of many Envirolet CT models in an outdoor setting.*

Pathogens thrive in excreta. The list includes E-coli, salmonella and the shigella bacteria, the parasite E. histolytica—the major cause of amoebic dysentery—and viruses. If the almost extinct polio virus again becomes a threat in developed countries, untreated sewage will be the primary means of transmission.

In the developing world, CT compost ("humanure") is applied directly to crop and pasture lands on the assumption that improved agricultural yields are worth the risk. Indeed, the modern composting toilet traces its ancestry to efforts by the North Vietnamese government to end the practice of using raw sewage as fertilizer.

U.S. authorities require that CT compost be either buried or disposed of in septic tanks or drain fields. Use on croplands is prohibited, although Oregon does make an exception for fruit trees. But not everyone is so cautious. Joe Jenkins, author of *Humanure: A Guide to Composting,* the bible of composting, asks:

If raw animal manure is considered safe for use on food crops, why not use treated compost?

If your family is healthy, how can its waste be infected?

However you opt to dispose of the material, wear long sleeves, gloves and a respirator when handling it. If the compost is dry and dusty, wet it down.

Design Requirements

Outdoor CTs are mounted high to be clear of flooding. Indoor installations are built up against an exterior wall on the sunny side of the house with the holding vault, or reactor, accessible from outside. The drawing in Fig. 2-9 shows the general arrangement for an interior, two-vault CT. When one vault fills, the toilet seat is moved over the other.

Brick or concrete vaults are sized to hold twelve months or more of compost. Each vault has its own hatch, which opens level with the vault floor. Steel hatches keep rodents out and transfer solar heat to the compost. The superstructure—the box-like platform that supports the toilet seat—can be made of plywood or whatever material is at hand.

Vents, usually made of PVC pipe, exhaust odor, CO_2, and water vapor to the outside and channel fresh air over the compost pile. The Jonah House CT follows good design practice by venting the pile from below with a series of drilled and straw-covered PVC pipes. Vents are screened. A tee on the discharge end of the pipe protects the compost from rain and creates a small draft on windy days. A wind-driven or electric fan can be installed at the vent outlet to improve circulation and reduce odor.

Caution: *The fan motor must be certified as explosion proof. Composting releases methane gas.*

The moisture in urine can slow the reaction enough to generate odor. Outdoor CTs sometimes drain urine out of the bottom of the vault; indoor toilets incorporate a baffle under the toilet seat. The urine goes into a sandy leach field or into a storage tank. When diluted with water, urine makes

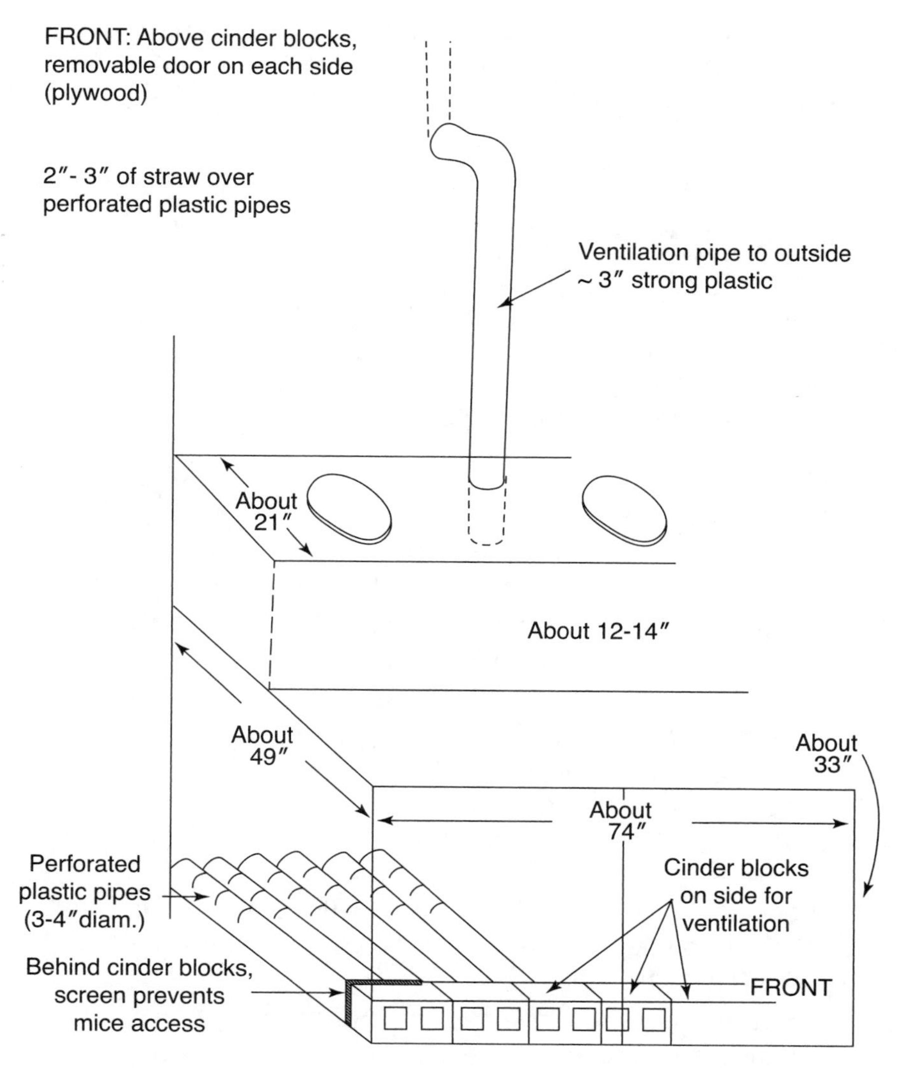

Figure 2-9. *A sketch of the Jonah House double-vault toilet. An especially noteworthy feature is the use of perforated PVC piping to aerate the compost from below. A baffle under the seat diverts urine through a hose and into a sealed drum on the outside of the building.*

excellent fertilizer for noncrop plants and trees. According to a Coalition Clean Baltic report, Swedish health authorities permit diluted urine that has been stored for six months to be used on crops.

CT Project Notes

Building a CT is a fairly obvious exercise; the sketch shown in Fig. 2-9 can serve as a guide. But fabrication of the PVC urine tray, which involved some unusual techniques, seems worth describing.

Parts needed to fabricate the urine tray:

One 3-in. PVC pipe, 10-in. long, cut in half lengthwise.

One 3-in. PVC pipe cap, cut in half.

Two ½-in. PVC male thread × slip couplings. One of these couplings is modified as described in Step 3.

One ½-in. insulating bushing (a PVC electrical conduit nut available as Rigid PN 47321).

One ½-in. female thread × ½-in. barbed hose connector (Apollo PN PXFAE1212).

Three feet of ⅜-in. vinyl tubing.

Step 1. Using a hacksaw, cut the 10-in. section of PVC pipe in half.

Step 2. Cut the 3-in. PVC cap in half. Drill a small pilot hole through the pipe and cap where the drain will be fitted (Fig. 2-10).

Step 3. Cut off the "slip" end of one ½-in. male thread × female slip coupling. Make the cut across the wrench flats to leave a flange about 1/16-in. thick on the threaded part (Fig. 2-11). Sand the cut smooth, but do not attempt to replicate the taper shown in the photo. That will come about later by applying heat during a trial assembly.

Step 4. Using the pilot hole as a guide, drill a ¾-in. hole in the pipe cap. The threaded end of the coupling will slip into the hole with only light sanding.

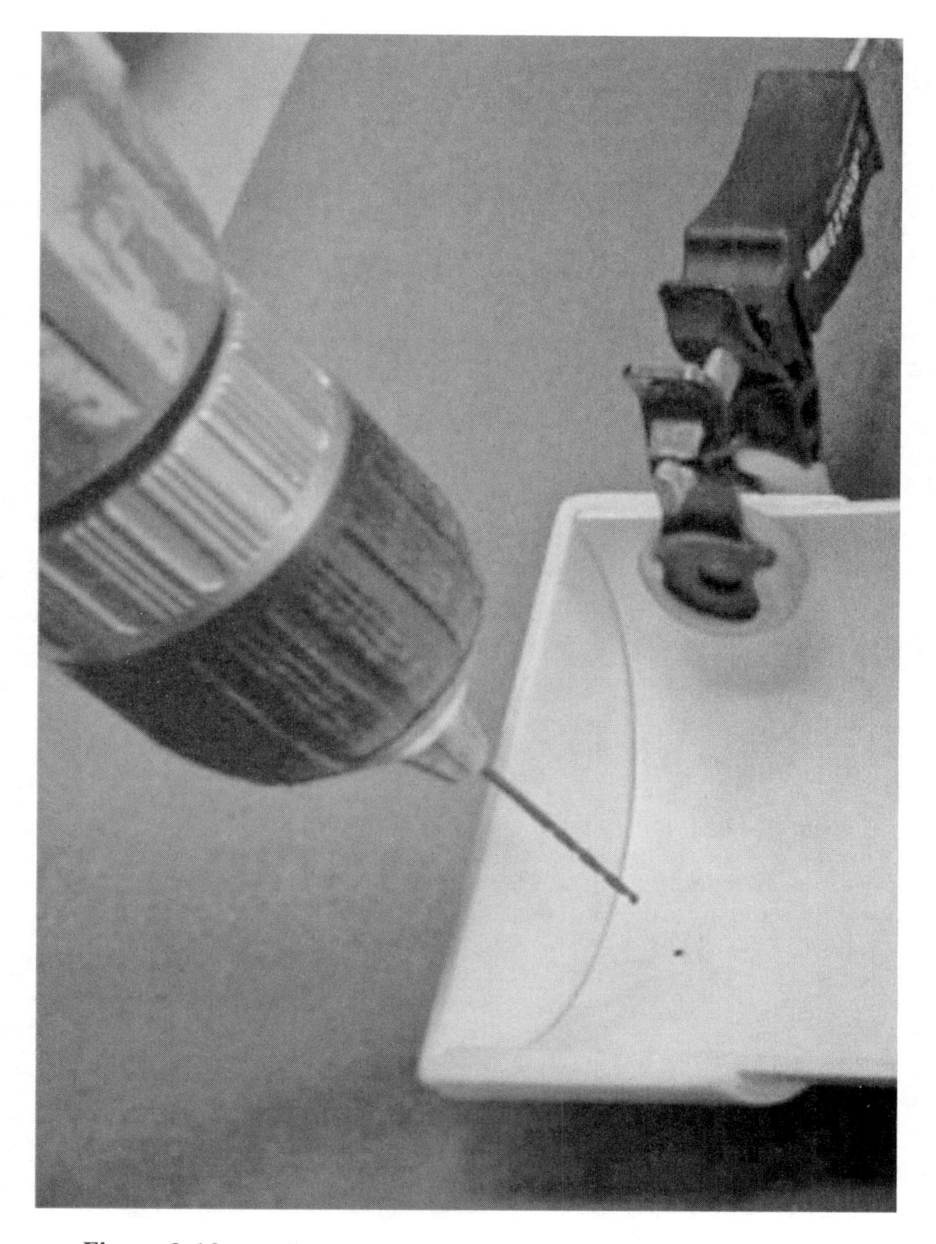

Figure 2-10. *Drilling a pilot hole for the drain through the end cap and pipe.* Tony Shelby

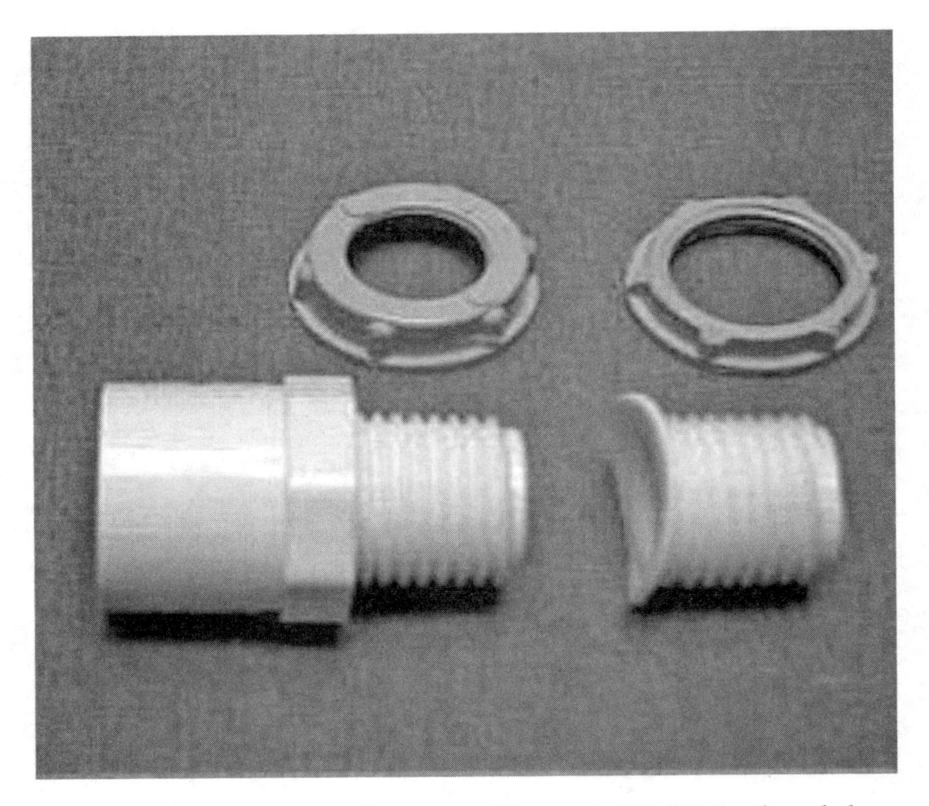

Figure 2-11. *Urine tray drain parts. The unmodified ½-in. threaded male × slip coupling is shown on the left. A second ½-in. male threaded × slip coupling yielded the part on the right. Note the flange that forms the seal against the round PVC pipe cap. The two nuts are bushings for PVC electrical conduit.* Tony Shelby

Step 5. Drill a 1-in. hole in the pipe to accept the flange on the modified coupling. Some sanding will be necessary (Fig. 2-12).

Step 6. With the modified fitting pressed into the hole in the cap, tighten the nut. Heat the top of the coupling (the flange) to conform to the arc of the pipe cap.

Step 7. Prime and apply a ⅛-in. bead of solvent cement to the pipe cap and flange. Insert the coupling and wipe off any solvent on the threads. Tighten the nut (Fig. 2-13).

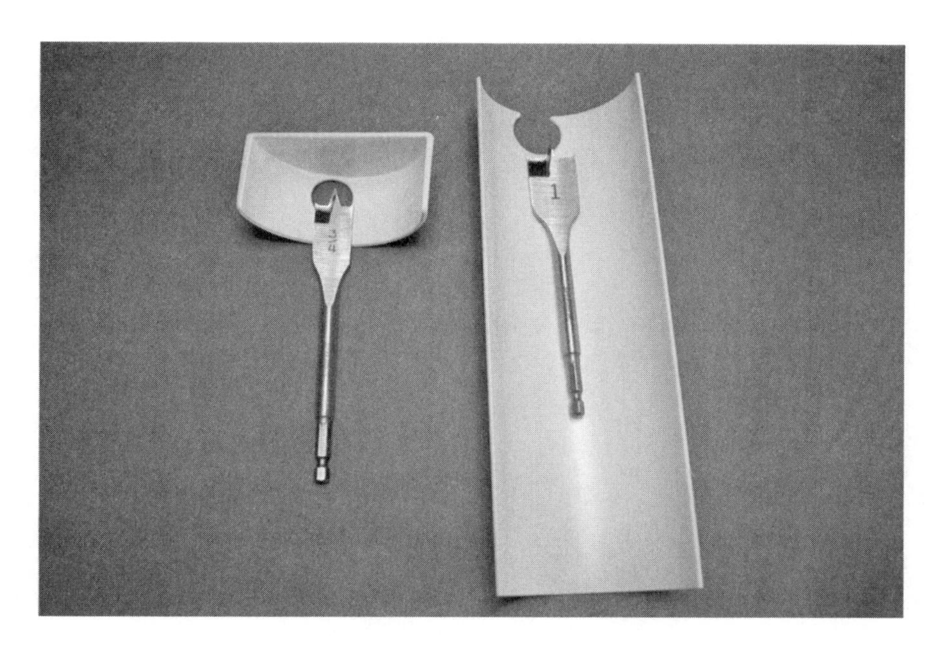

Figure 2-12. *Tray drain holes.* Tony Shelby

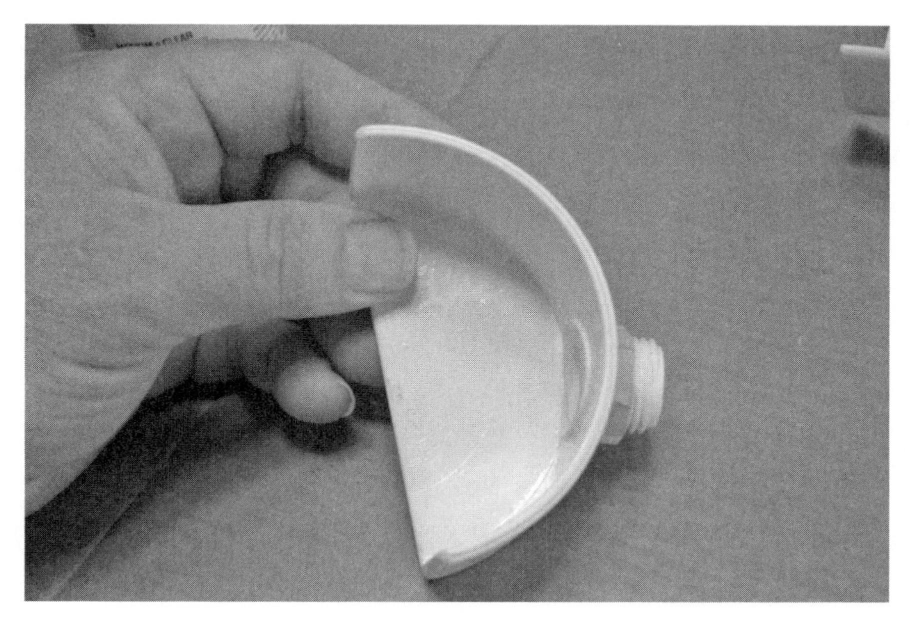

Figure 2-13. *One element of the urine tray drain in place.* Tony Shelby

Figure 2-14. *Tray glued and clamped.* Tony Shelby

Step 8. Solvent-weld the pipe section to the end caps. Press the parts together and clamp (Fig. 2-14). Make up the tray drain (Fig. 2-15).

Troubleshooting CTs

The most common complaints are:

- **Sewage odor**—usually caused by excessive moisture. Mix dead leaves or other dry organic material into the compost.
- **Slow decomposition**—composting goes best with about 40 percent moisture, or the degree of moisture left on a well wrung-out dishcloth. Urine or green materials, such as freshly cut grass clippings, should restore the balance.

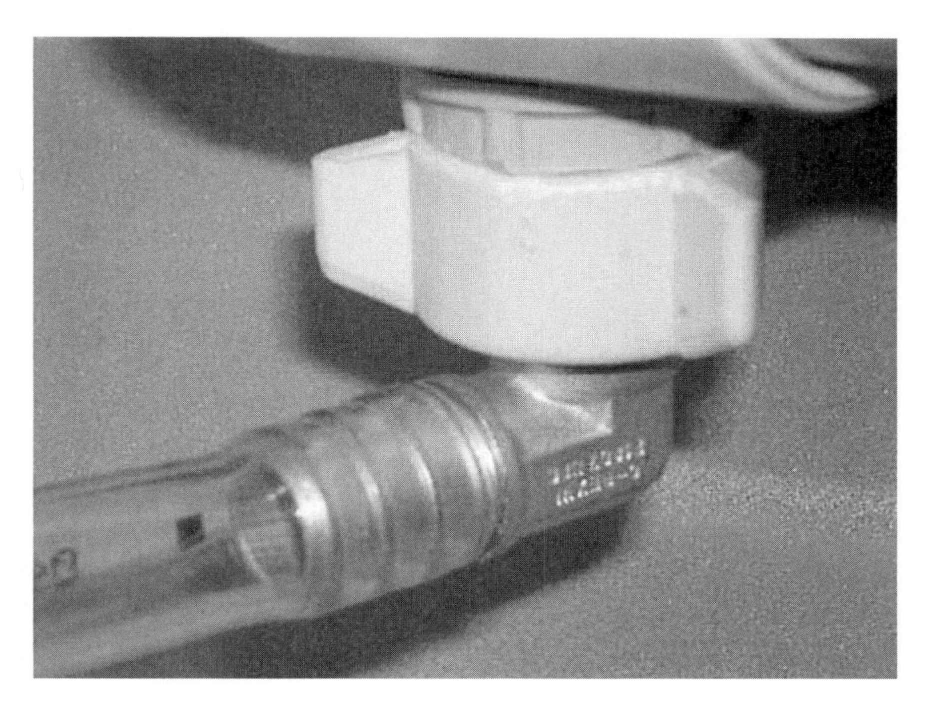

Figure 2-15. *The urine tray drain installed.* Tony Shelby

- **Ammonia odor**—the sign of excessive nitrogen. For composting to work, the carbon-to-nitrogen ratio should range from between 25 and 30 parts carbon to one part nitrogen. Raise the carbon ratio with a brown filler, such as shredded newspaper, dry leaves, or wood shavings.

3

Water from Air

This chapter describes how to harvest rain and the condensate that air conditioners produce.

Rainwater Harvesting (RWH)

Since agriculture began, people harvested rainwater with artificial ponds, ditches, berms, and terraced hillsides. The idea that lawns and ornamental gardens should be planted dead flat originated with the Victorian passion for regularizing nature.

More elaborate systems consisted of a catchment area and a means of storing the runoff for later use.

Calculations

Yield:

- Catchment area for rectangular roofs = Width × Length. A 40-ft by 50-ft roof has a catchment area of 2000 ft^2.
- 1 inch of rain deposits 0.62 gal/ft^2.
- Yield = inches of precipitation per month × 0.62 × catchment square footage.
- A 2000 ft^2 catchment area should yield 1240 gal per in. of rainfall.

Demand:

- Lawn and garden irrigation accounts for about 60 percent of household water use, and 80 percent or more in dry areas.
- Toilets are typically flushed seven times a day by each

resident. Older toilets averaged around five gallons per flush, or 1050 gal per person per month. Newer Energy-Sense toilets use 1.28 gpf.

- Each household resident uses about 50 gallons per day, or 1500 gal/month.

Supply:

- Supply = monthly yield minus monthly demand.
- Size the holding tank for a three-month dry spell. If rainwater consumption averages 900 gal./month, the tank should hold 2700 gal.

Tank Volume:

- Cylindrical tanks Radius (ft) × Radius (ft) × Height (ft) × 3.14 = cubic ft.
- Rectangular tanks Height (ft) × Width (ft) × Depth (ft) = cubic ft.
- Cubic ft × 7.48 = gallons.

This approach to rainwater harvesting was used in China and in the Middle East as early as 6000 years ago. Some of the best-documented examples supplied water to the vast urban complexes built by the Maya AD 250–900 in the Yucatan and Central America. These cities housed upwards of 100,000 people and included temples, palaces, and pyramids, several of them larger than those in Egypt. Some of these complexes were built on riverbanks; others depended upon shallow wells for their water needs. But many survived and prospered entirely upon rainwater. Large paved plazas collected runoff from building roofs and diverted it to underground cisterns. The water was then used for irrigation and, with filtering, for human consumption.

One can still encounter abandoned cisterns in rural America, mementos of a time when farm families supplemented hand-drilled wells with rainwater. Until the advent of municipal water systems, town dwellers did the same.

Modern Examples

Until recently, RWH has been considered an obsolete technology, as something we in our great wisdom have progressed beyond. But this attitude is changing, and never more dramatically than in the developing world, where NGOs and governmental agencies look to rainwater as a means of reducing disease, improving crop fertility, controlling runoff, and surviving droughts. A program in drought-stricken northeastern

Brazil aims to install one million cisterns, and, since 1986, more than two million catchment systems have gone into operation in the Gansu province in China. Similar efforts are underway in the Philippines and across much of Africa. Bermuda and several other Caribbean islands require that all new construction include a provision for a self-sustaining water supply system, either by capturing rainwater or drilling a well.

In the United States, cost considerations rather than public health concerns drive the change. The cost of municipally supplied water has increased by almost by a third during the past five years. And costs will continue to escalate as treatment systems revamp to meet Safe Drinking Water Act mandates. Since most U.S. treatment infrastructure was installed after World War II, it is long past its 50-year design life. Add to this situation the need to enlarge treatment and distribution facilities to meet the growth in population, and one can understand the emphasis governmental authorities put on RWH.

Consequently, RWH approaches the norm for new governmental and commercial construction. It is estimated that some 100,000 professionally engineered RWH installations are in operation. In addition to reducing stress on municipal supplies, these installations save their owners money. The harvesting system installed at the Advanced Micro Devices semiconductor fabrication plant in Austin, Texas, reduced the company's water bills by $1.5 million per year. The City of North Miami Beach invested $30,000 in a catchment system that yields 30,000 to 40,000 gallons a month for irrigation and bulk truck needs. Maintenance costs have been minimal.

Similar efforts are underway in other developed countries. One interesting example is Singapore's Changi Airport. Water runoff from the runways is stored underground and used for toilet flushing and fire-fighting drills.

Homeowner Systems

What follows deals with homeowner systems intended primarily for yard and garden irrigation. The nonpotable water produced by these systems can also be used for car washing and other outdoor purposes. Having a thousand or more gallons of water on hand can be critical for fire fighting in rural areas. Agricultural water can also be piped indoors for toilet flushing and clothes washing.

Making collected rainwater potable, that is, safe to drink, bathe in, and prepare food with, entails health risks, at least in the context of what authorities in developed countries consider tolerable. And the experts do not always agree on the best ways to reduce these risks. See "Potable Rainwater" on pages 76–82 for a discussion of risks and treatment options.

While potable RWH is controversial, capturing rain for irrigation and other outdoor uses is a win-win situation. The collected water has the following advantages:

- It costs nothing, once the collection system is up and running.
- It contains no salt and none of the chlorine-induced carcinogens characteristic of municipal water supplies. Plants thrive on it.
- It relieves stress on municipal systems, many of which are under-funded and poorly maintained.
- It reduces flooding and erosion by holding water on site. Some fraction of the collected water then finds its way into the water table. One of our nightmares is the disappearance of groundwater.

Basic systems require no pumps and little filtration. If we want to use the water for toilet flushing and clothes washing, more by way of filtration and treatment is recommended. Indoor systems also require dedicated plumbing and a pressure tank.

Getting Started

Begin by surveying the grounds with an eye to water flow. Any slope in the ground, assisted by sidewalks, ditches, or buried PVC piping, can direct water from rooftops and driveways to garden areas. In order to hold water and give it time to percolate to root depth, locate plants and trees in shallow depressions or in holding basins created by earthen walls, or berms. Large basins, extending beyond the "drip line" (the area under the leaves) encourages roots to spread out laterally. The wider the reach of root systems, the better the chances plants have of surviving drought.

Rain Barrels

Once you have taken advantage of the topography, the next step is to experiment with one or more rain barrels. Figure

Figure 3.1. *Rain barrel placement requires a bit of thought.*
City of San Diego

3-1 provides some useful tips on siting rain barrels around the property. Barrels should hold at least 55 gallons, have a tightly fitting lid, and have a bib threaded for a garden hose. Commercial rain barrels, made from recycled food-grade plastic, come with all the fittings, but cost upwards of $150. Wooden barrels cost even more. Some DIYers use corrugated storm conduit (Fig. 3-2).

After looking at the options, we settled on 55-gal food barrels purchased from a restaurant for $20 apiece. Some restaurants give the barrels away, since they cannot be reused for food storage. The following table lists the cost and materials required for each barrel.

Purchased Components	**Cost ($)**
1–recycled food 55-gal food barrel	20.00
2–bulkhead fittings @ $10 each	20.00
1–hose bib, brass	8.00
1–type A gutter downspout ell	2.85
1–3-in. × 4-in. PVC slip-fit coupling	6.60
1–¾-in. female thread × 1-in. slip-fit PVC coupling	1.05
1–roll Teflon tape	4.10
1–Handy Pack, with PVC primer and cement	7.47
4–8-in. × 8-in. concrete blocks @ $1.08 each	4.32
Total	$74.39

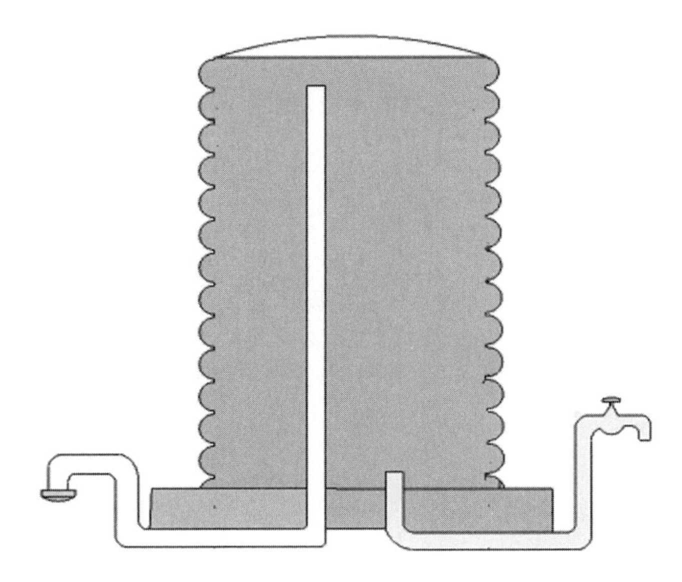

Figure 3-2. *A 7- or 8-ft. length of 3½-ft corrugated steel conduit makes an inexpensive, though ugly container. A 4 ft by 4 ft × 10-in. reinforced concrete pad, supported on gravel or bricks, seals the bottom of the conduit and anchors the 1-in. PVC overflow (shown on the left) and the ¾-in. PVC discharge lines. Screen wire, looped over the top of the pipe and secured with wire or bungee cords, keeps out insects. If you intend to drink the water, coat the inside of the conduit with a food-grade sealant.*

An aquarium supply house supplied the bulkhead fittings; other components came from big-box stores or from the scrap box. Once materials were in hand, the project required about two hours to complete.

Step 1. Using a 1¾-in. hole saw, drill the hole for the hose bib (Fig. 3-3). Note that the bib should be several inches above the bottom to leave space for sediment.

Step 2. Drill a second 1¾-in. hole near the top of the barrel for the overflow (Fig. 3-4).

Step 3. Install the bulkhead fitting for the hose bib. Lay the barrel on its side and, while an assistant holds the fitting

Figure 3-3. *The hose bib mounts in a $1\frac{3}{4}$-in. hole near the bottom of the barrel.* Tony Shelby

Figure 3-4. *Drilling the overflow hole, which should be as high on the barrel as practical.* Tony Shelby

in place, reach inside the barrel to make up and tighten the fitting nut. Do not overtighten—we are working with plastic.

Step 4. Install the overflow bulkhead fitting (Fig. 3-5).

Step 5. Set the barrel upright. Apply PVC primer and adhesive to the slip portion of the ¾-in. female thread × 1-in. slip coupling. Insert the coupling into the lower bulkhead fitting. Hold it in place until the adhesive dries.

Step 6. Wrap the brass hose bib thread with three or four turns of Teflon tape and assemble the bib to the bushing (Fig. 3-6). Make up the connection moderately tight with the bib nozzle pointing down.

Step 7. Overflow from the barrel should be directed away from the house. We constructed a screened outlet, as shown in Figure 3-7.

Figure 3-5. *Once the hose bib is secured, install the overflow bulkhead fitting.* Tony Shelby

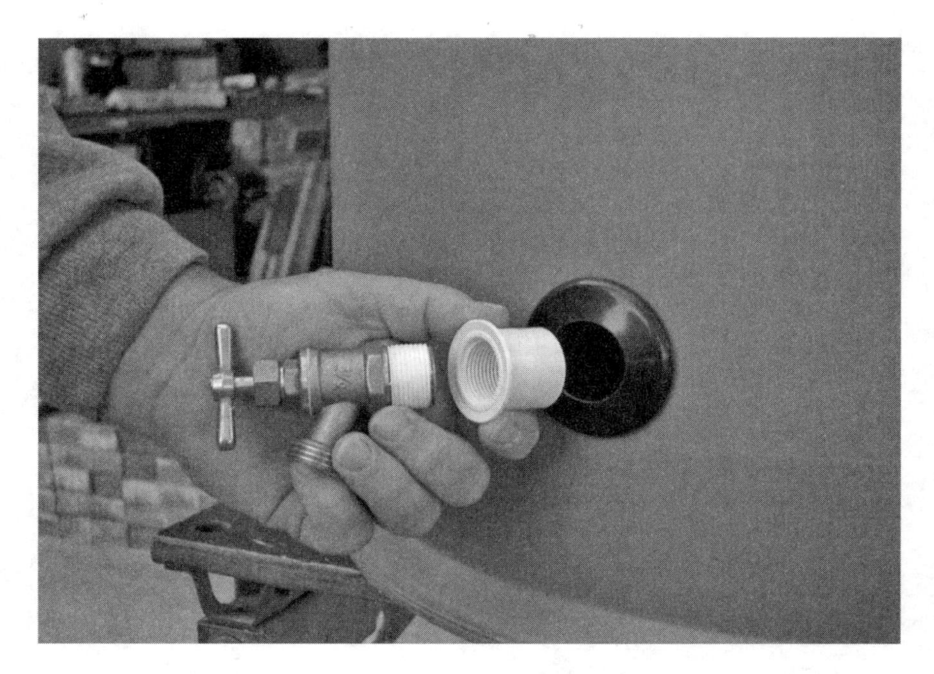

Figure 3-6. *Hose bib, PVC coupling, and bulkhead.* Tony Shelby

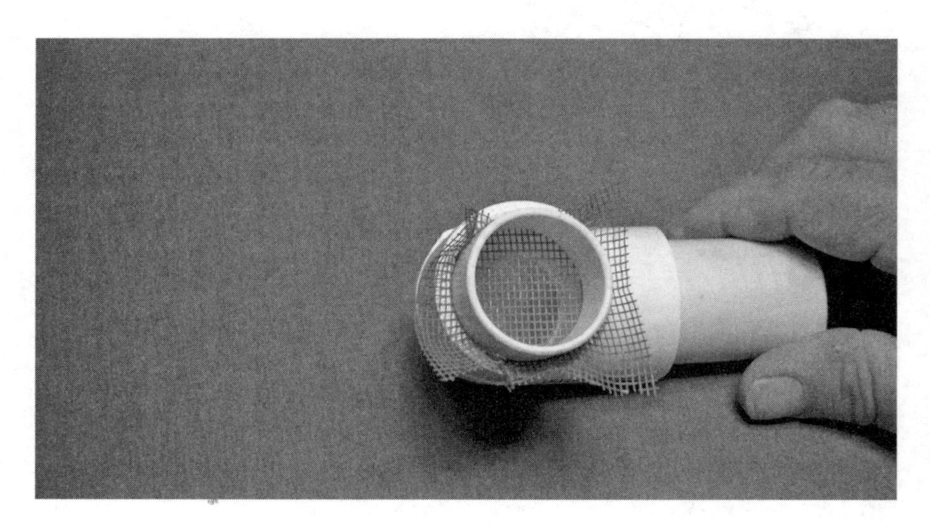

Figure 3-7. *Overflow outlet constructed from a short piece of 1-in. PVC and a 90° ell. A second piece of 1-in. PVC wedges the screen in place and, because it is not glued, it can be removed for maintenance.* Tony Shelby

Step 8. Cut a 3-in. disc from PVC and, using it as a template, make a matching hole in the lid (Fig. 3-8). Large-diameter PVC drain pipe is the source for flat panels. Cut out the panel and, using a heat gun, heat the material and press it flat between the boards.

Warning: *Work in a well-ventilated area and do not overheat to the degree that the plastic changes color. PVC fumes contain dioxin.*

Step 9. Solvent weld the PVC disc into the 3-in. × 4-in. slip-fit coupling.

Figure 3-8. *Use a jigsaw to cut the 3-in. hole for the inlet fitting. Note that the hole is offset from the center of the lid.* Tony Shelby

Step 10. Cut a piece of the screen wire and place it over the coupling (Fig. 3-9). A short length of 4-in. PVC pressed into the adapter secures the screen.

Step 11. Insert the coupling into the hole previously cut in the lid. The coupling will rest on its shoulder with its large end up.

Step 12. Place the barrel on level ground, supporting it with concrete blocks. Raising the barrel a few inches enables a watering can to fit under the hose bib.

Step 13. Cut the downspout to length and splay it so that it makes up to the inlet coupling (Fig. 3-10). We used two

Figure 3-9. *A piece of 4-in. PVC pipe, pressed into—but not solvent-welded—the inlet coupling secures the screen.*
Tony Shelby

Figure 3-10. *The downspout must be modified slightly to fit.*
Tony Shelby

downspout elbows, one pointing away from the house and one pointing down, to align with the barrel inlet.

Figure 3-11 shows the finished barrel. Pleased with the project, we then installed a second barrel (Fig. 3-12).

Dedicated Irrigation Systems

The first consideration is to determine what, if any, legal constraints exist in your area. Several Western states prohibit or severely curtail rainwater harvesting because of laws dating back to the nineteenth century, when rainwater was considered a zero-sum game. What one person collected, another lost. Colorado laws technically prohibit rain barrels in areas served by municipal water. But change is underway: Utah recently amended its legislation to permit RWH, and, as of

Figure 3-11. *The finished rain barrel.* Tony Shelby

Figure 3-12. *Subsequently we installed a second barrel and painted both*. Tony Shelby

June 1, 2010, the City of Tucson requires that 50 percent of a commercial property's irrigation water must be supplied by rain. Texas will defray as much as $40,000 of system costs.

The website http://www.harvesth2o.com/statues_regulations.shtml provides a fairly comprehensive overview of U.S. and Canadian legislation in this area, but check with city and county authorities. RWH may also conflict with deed restrictions and with rules imposed by homeowners' associations. In most jurisdictions, anything more elaborate than placing rain barrels under existing downspouts involves plumbing and/or building permits.

System Potential

Once you've determined the legalities, try to get a sense of the scale of the project, of how much water you can reasonably expect to capture. Yield depends upon the amount of precipitation and the surface area of the collector, which is nearly always a rooftop. The uncertainties of rain and climate change complicate the picture. But in the absence of anything better, historical data is all we have to go by.

Table 3-1 provides 30-year precipitation data broken down by month for the lower-48 states. City governments, county agents, or agricultural extension offices can provide local data. Readers in the American Southwest might want to look at long-term drought predictions for the area.

Figure 3-13 illustrates roof footprints. To calculate the catchment area multiply the width in feet of the footprint times its length. As a rule of thumb, allowing for spillover from flooded gutters, leaks, and water that has been absorbed by common roofing materials, each square foot (ft^2) of surface area collects 0.62 gal per inch of rainfall. One inch of rain on a 1000 ft^2 roof should, by this conservative estimate, yield 620 gallons.

$$\text{Annual yield (gal)} = \text{collection area (ft}^2) \times 0.62 \times \text{precipitation (in./year)}$$

Metal and tile roofs do better, with theoretical yields as high as 0.80 gal. per ft^2.

Table 3-1
Average Precipitation by State between 1971 and 2000

State	Jan	Feb	Mar	Apr	May	Jun	Jul	Aug	Sep	Oct	Nov	Dec	Total
Alabama	5.91	5.18	6.65	4.77	4.69	4.59	5.39	3.90	4.26	3.16	4.79	5.01	58.28
Arizona	1.30	1.25	1.35	0.54	0.42	0.31	1.74	2.09	1.30	1.20	0.97	1.13	13.61
Arkansas	3.65	3.61	4.93	4.82	5.20	4.31	3.52	2.87	3.72	4.08	5.34	4.72	50.78
California	4.14	3.95	3.55	1.40	0.83	0.32	0.19	0.30	0.58	1.20	2.63	3.11	22.20
Colorado	0.79	0.75	1.29	1.47	1.92	1.44	2.01	1.95	1.35	1.20	1.03	0.77	15.97
Connecticut	4.30	3.23	4.42	4.30	4.33	4.05	4.19	4.39	4.36	4.31	4.42	4.08	50.39
Delaware	3.92	4.34	3.51	4.18	3.52	4.05	4.63	4.19	3.38	3.31	3.55	3.55	45.68
Florida	3.61	4.13	2.84	3.92	6.94	7.19	7.26	6.39	3.46	2.88	2.76	3.17	54.57
Georgia	5.20	4.49	5.28	3.62	3.58	4.44	5.05	4.66	3.89	3.01	3.55	3.94	50.72
Idaho	2.12	1.72	1.73	1.56	1.97	1.51	0.92	0.84	1.08	1.30	2.07	2.14	18.96
Illinois	1.97	1.99	3.22	3.83	4.31	4.12	3.94	3.69	3.24	2.87	3.41	2.74	39.32
Indiana	2.44	2.31	3.40	3.96	4.47	4.25	4.20	3.88	3.18	2.92	3.62	3.09	41.72
Iowa	0.95	0.98	2.20	3.33	4.23	4.62	4.27	4.18	3.40	2.51	2.13	1.24	34.05
Kansas	0.77	0.91	2.25	2.59	4.18	3.88	3.61	3.16	2.68	2.18	1.73	0.98	28.92
Kentucky	3.77	3.77	4.60	4.19	5.05	4.34	4.47	3.71	3.49	3.13	4.09	4.38	48.98
Louisiana	5.91	4.66	5.32	4.89	5.46	5.20	5.15	4.47	4.55	4.05	5.06	5.38	60.09
Maine	3.46	2.52	3.37	3.46	3.64	3.77	3.74	3.61	3.68	3.70	3.82	3.51	42.28
Maryland	3.66	3.03	4.09	3.43	4.27	3.80	4.11	4.01	4.04	3.38	3.41	3.41	44.64

Table 3-1
Average Precipitation by State between 1971 and 2000 *(continued)*

State	Jan	Feb	Mar	Apr	May	Jun	Jul	Aug	Sep	Oct	Nov	Dec	Total
Massachusetts	4.17	3.34	4.16	4.11	3.91	3.84	3.83	4.02	3.94	4.16	4.35	4.05	47.88
Michigan	2.02	1.45	2.22	2.69	2.94	3.23	3.15	3.58	3.65	2.85	2.77	2.28	32.84
Minnesota	0.86	0.64	1.45	2.06	3.06	4.17	3.95	3.69	2.91	2.29	1.58	0.77	27.44
Mississippi	5.92	4.96	6.30	5.61	5.39	4.47	4.80	3.67	3.81	3.45	5.19	5.64	59.23
Missouri	1.95	2.15	3.51	4.04	4.82	4.26	3.96	3.58	3.97	3.36	3.82	2.81	42.23
Montana	0.80	0.60	0.90	1.30	2.34	2.40	1.65	1.35	1.33	1.05	0.86	0.80	15.37
Nebraska	0.51	0.60	0.66	2.38	3.75	3.45	3.27	2.63	2.10	1.56	1.16	0.57	23.63
Nevada	0.96	0.92	1.09	0.76	1.01	0.63	0.51	0.63	0.71	0.75	0.81	0.76	9.54
New Hampshire	3.42	2.62	3.37	3.50	3.76	3.85	3.94	3.97	3.66	3.95	3.93	3.44	43.42
New Jersey	3.94	2.96	4.18	3.93	4.28	3.77	4.46	4.57	4.14	3.51	3.70	3.71	47.15
New Mexico	0.67	0.60	0.71	0.63	1.11	1.23	2.27	2.64	1.82	1.34	0.83	0.78	14.63
New York	2.99	2.41	3.12	3.42	3.68	4.00	3.81	3.86	4.14	3.48	3.74	3.24	41.90
North Carolina	4.51	3.71	4.67	3.47	4.28	4.40	5.04	4.97	4.78	3.61	3.45	3.57	50.45
North Dakota	0.50	0.45	0.80	1.40	2.31	3.19	2.75	2.10	1.74	1.41	0.73	0.44	17.82
Ohio	2.51	2.27	3.06	3.47	4.06	4.10	4.09	3.77	3,10	2.62	3.16	2.93	39.16
Oklahoma	1.48	1.78	3.07	3.32	5.13	4.24	2.73	2.75	3.80	3.39	2.79	2.05	36.55
Oregon	3.79	3.16	2.93	2.18	1.88	1.27	0.62	0.73	1.07	1.90	3.97	4.06	27.55
Pennsylvania	3.04	2.58	3.41	3.55	4.07	4.45	4.15	3.78	4.08	3.18	3.58	3.16	43.02

Table 3-1
Average Precipitation by State between 1971 and 2000 *(continued)*

State	Jan	Feb	Mar	Apr	May	Jun	Jul	Aug	Sep	Oct	Nov	Dec	Total
Rhode Island	4.45	3.63	4.65	4.32	3.72	3.52	3.20	3.99	3.80	3.79	4.55	4.36	47.98
South Carolina	4.70	3.79	4.64	3.16	3.69	4.72	5.05	5.29	4.51	3.42	3.21	3.67	49.84
South Dakota	0.45	0.53	1.30	2.11	3.10	3.24	2.75	2.13	1.66	1.62	0.82	0.44	20.14
Tennessee	4.65	4.31	5.60	4.51	5.33	4.48	4.67	3,52	3.88	3.32	4.77	5.18	54.22
Texas	1.59	1.67	1.85	2.14	3.52	3.35	2.22	2.54	3.19	2.95	2.02	1.84	28.87
Utah	1.05	0.99	1.21	1.07	1.20	0.64	0.88	1.01	1.07	1.30	0.99	0.82	12.26
Vermont	3.11	2.33	3.07	3.28	3.84	3.93	4.21	4.49	3.97	3.67	3.73	3.19	42.82
Virginia	3.64	3.13	4.04	3.42	4.26	3.79	4.34	3.84	4.00	3.50	3.29	3.15	44.39
Washington	5.38	4.37	3.76	2.71	2.23	1.77	1.02	1.05	1.65	3.05	5.85	5.93	38.78
West Virginia	3.46	3.11	3.97	3.62	4.57	4.23	4.75	4.13	3.51	3.05	3.46	3.43	45.30
Wisconsin	1.22	1.00	1.96	2.86	3.37	4.02	4.07	4.27	3.74	2.50	2.29	1.35	32.64
Wyoming	0.63	0.57	0.86	1.36	2.06	1.58	1.33	1.02	1.14	1.04	0.76	0.62	12.97

(Calculated by NOAA-CIRES from data obtained by the National Climatic Data Center)

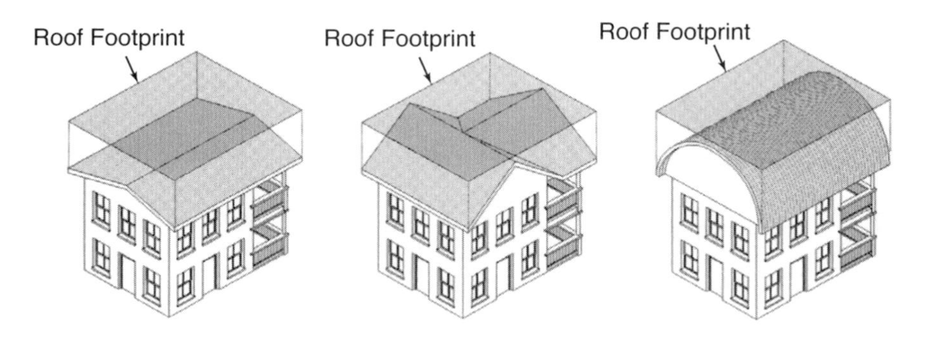

Figure 3-13. *Footprints for rectangular roofs. For round roofs, multiply radius times radius times 3.14.* Texas Water Development Board

Storage Tanks

Rather than harvest every drop that falls, many homeowners let their budgets determine tank size. Better utilization comes about if we go slow and view the system as a work in progress, adding storage and collection surface as experience indicates. If the existing tank chronically overflows, install another. If the tank runs dry between rains, consider ways to increase the size of the collection area. Driveways and sidewalks can be pressed into service, although a small pump will be needed to convey the runoff to a surface tank.

Contractors usually recommend that tanks be sized to provide enough water to compensate for three months without rain. In drought-prone areas, storage should be larger. On the other hand, and there is nearly always another hand when dealing with RWH, the British Standards Institute (BSI) limits tank size to five percent of the system's annual yield. Five percent of yield can mean as little as three weeks of reserve capacity. The reason for setting limitations is to avoid problems associated with stagnation. Stagnant water is a breeding ground for microorganisms, including the one that causes Legionnaire's Disease.

Underground tanks, often called cisterns, add cost and complexity to the project (Fig. 3-14). Most installations have factory-supplied surface tanks that can be hidden in base-

Figure 3-14. *An underground tank under construction by the Sparkle Tap Water Co.* Jack Holmgreen

ments or obscured with trestles. And they need not have an industrial look (Fig. 3-15).

Tanks should incorporate one or more screened vents, inlet, outlet, and overflow fittings, and a manhole for access. Some include a trapdoor at the bottom to assist in removing sediment. You might also consider a partition for large, site-built tanks, to enable each half of the tank to be decommissioned for cleaning. A sight glass or a float gauge is almost mandatory.

Food-grade polyethylene tanks, available in surface-mounted versions and with reinforcing for subterranean installation, cost $0.50 to $0.80 per gallon. Welded steel tanks are the most expensive with costs ranging from $2 to $2.50 per gallon. The larger the tank, the lower the per-gallon cost.

Plastic tanks should be labeled as meeting ANSI/NSF Standard 61 and/or U.S. FDA regulation 21CFR 177.1520 (1) 3.1 and 3.2 for resins. The National Resource Conservation Service imposes construction standards on tanks used for animal watering. A two-year warranty is standard, although

Figure 3-15. *A water tank disguised as a giant ladybug in Round Rock, Texas.*

several manufacturers offer 10-year warranties at extra cost. Table 3-2 lists other options. If money is no object, stainless steel would be an excellent choice.

The tank must rest on a reasonably firm, stable foundation. Swelling clays invite early failure. Sand makes an ideal interface, since it conforms to irregularities on the tank bottom and, when banded with concrete retaining walls, easily supports the weight (Fig. 3-16). A 1000-gal tank holds 8350 lb of water.

Jack Holmgreen, owner of the Houston-based Sparkle Tap Water Co., installed the RWH system pictured here. In order

Table 3-2
Tank Materials

Material	Advantages	Disadvantages
Polyethylene	Cost-effective and can be used above or, with suitable reinforcement, below ground.	Ultraviolet (UV) degradable. Must be painted or tinted, and placed on a stable foundation.
Fiberglass	Same as polyethylene.	Brittle, fatigue-prone. Interiors must be lined or coated with a food-grade sealant
Stainless steel	Virtually indestructible, easy to mount.	Expensive.
Galvanized steel	Readily available.	Limited to above-ground installation. Interiors must be lined or coated with a food-grade sealant..
Sprayed-on ferro-concrete, e.g., Gunite	Durable for above- or below-ground installation. Fabricated on site.	Immoveable and prone to cracking. Requires special skills to construct.
Poured-in place concrete	Same as ferro-concrete.	Same as sprayed-on-ferro-concrete
Concrete block or stone	Durable, fabricated on site.	Requires masonry skills to construct and repair.
Cypress and other durable woods	Aesthetic, fabricated on site.	High labor and materials costs.

Figure 3-16. *A sand-filled pad, contained within a concrete form, for a 1000-gal polypropylene tank.* Tony Shelby

not to disturb the sediment layer, Jack brings the Schedule (Sch.) 40 PVC inlet pipe in low and fits it with an upward pointing ell and a turbulence-reducing calming nozzle (Fig. 3-17). The discharge pipe mounts a few inches above the bottom of the tank to allow sediment to build. Had a pump been fitted, it would draw from a floating inlet hose (Fig 3-18). PVC plumbing exposed to sunlight was painted.

Help will be needed to unload and maneuver the tank into position (Fig. 3-19). Once on the pad, hook-up proceeds quickly (Fig. 3-20).

Most installers do not paint polyethylene tanks, which are tinted at the factory for UV light protection. Jack disagrees and applies two coats of inexpensive latex or elastomeric paint to tank exteriors. Either type of paint is flexible enough not to crack as the tank expands and contracts with temperature changes. A coat of black paint goes on first to darken the interior. Algae must have light to grow. A green second coat reduces water heating in warm climates.

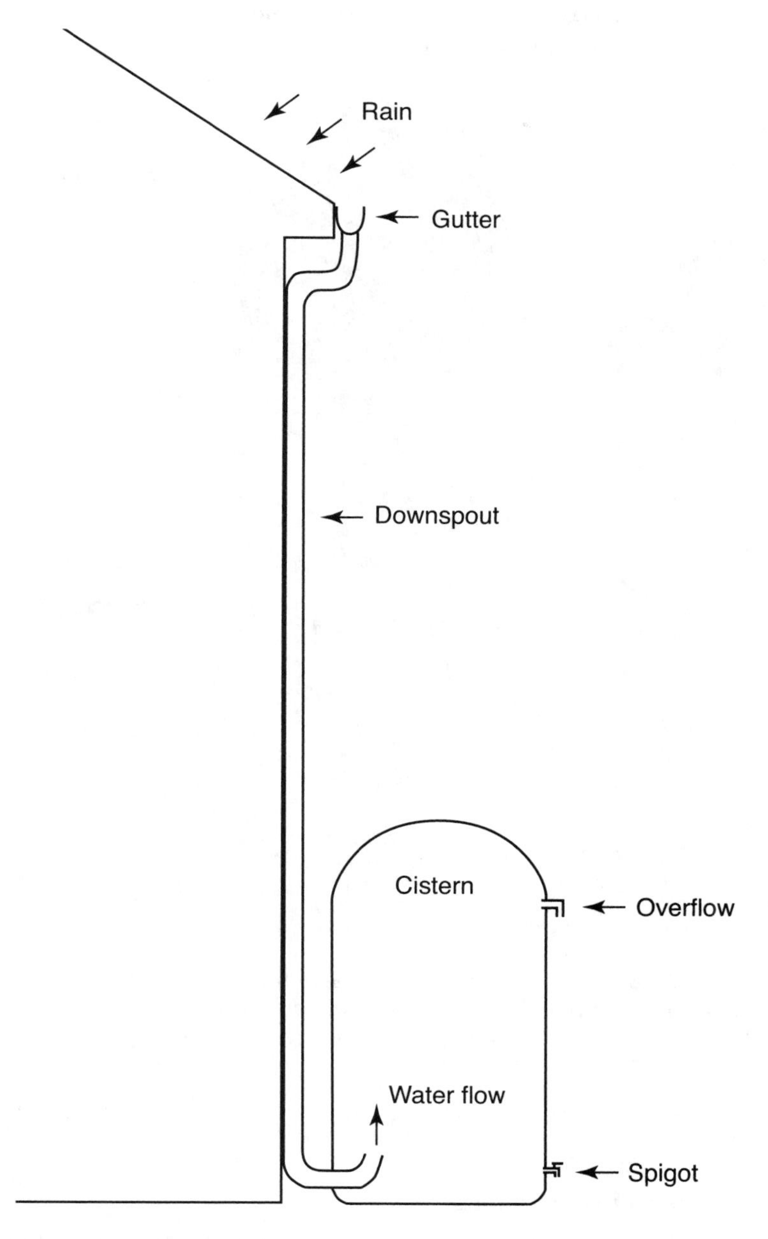

Figure 3-17. *A Jack Holmgreen installation with a gravity-fed outlet. Jack mounts the upward-pointing inlet low with the discharge passing through a calming nozzle. Other installers pipe water in from the top of the tank.* Tony Shelby

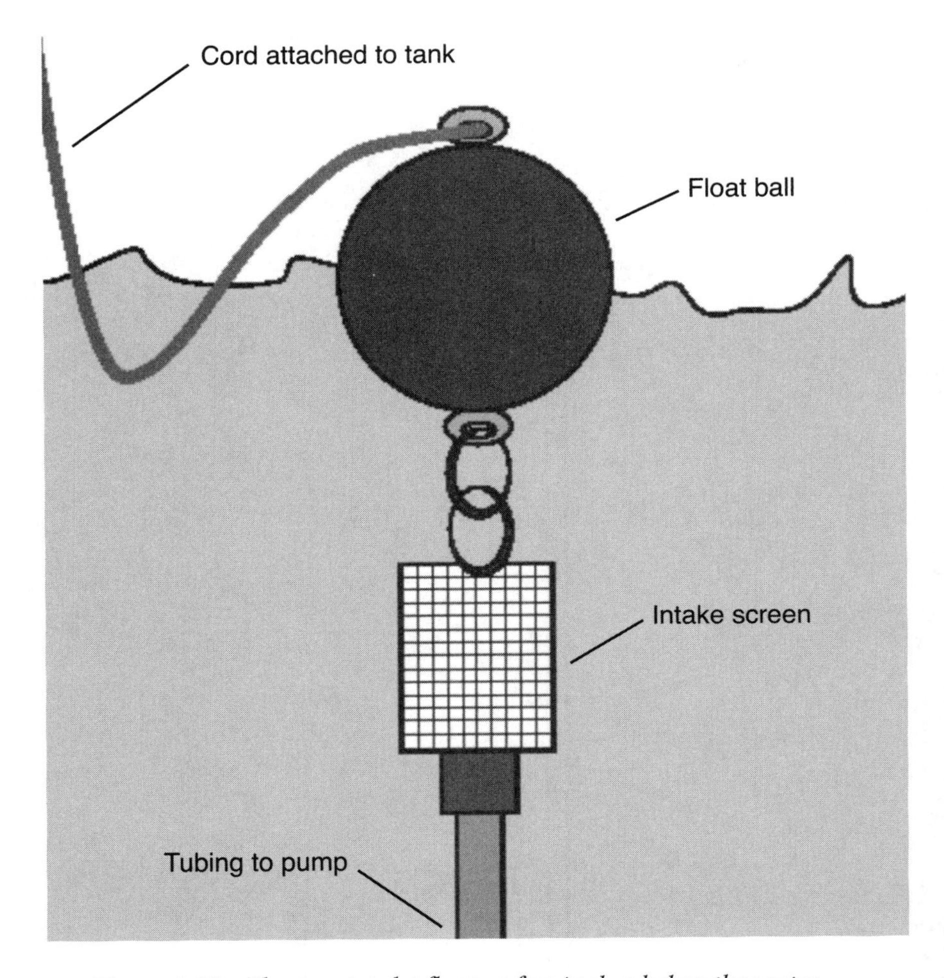

Figure 3-18. *The pump inlet floats a few inches below the water level and incorporates a screen or, when used for potable water, a filter.* Oregon Dept. of Consumer and Business Services

Gutters and Downspouts

Study the roof configuration, noting the slope and the valleys where gabled roofs intersect to determine if additional gutters and downspouts should be installed. Follow the roof edge

Figure 3-19. *Poly tanks weigh little, but are awkward to handle.*
Tony Shelby

around the structure and record the location of existing and planned downspouts with a view to minimizing pipe runs to the storage tank. A dimensioned sketch will enable you to make a reasonable estimate of the cost.

Installers generally prefer 0.025 in. or thicker aluminum gutters and downspouts. Gutters should be

- 5-in. wide with a slight downward slope and their outer edges 1-in. lower than their inboard edges,
- fitted with expansion joints between long runs,
- provided with hangers at 3-ft intervals, and
- fitted with leaf screens.

Downspouts should be spaced at 20- to 40-ft intervals and sized to provide a drain area of 1 in.^{2} per 100 ft^2 of collector surface. Use Schedule 40 PVC for transfer to the holding tank.

Figure 3-20. *Once the tank is in position, hookup is almost anticlimactic*. Tony Shelby

Filtration

A stainless steel or nylon screen, either at the tank inlet or on the downspout, keeps leaves and other organic material out of the tank. The screen should be accessible for cleaning. Well-designed irrigation systems and those that supply potable water also incorporate a first-flush diverter. Figure 3-21 illustrates one version of the device, which consists of a horizontal pipe "teed" to a standpipe. Initial flow from the roof collects in the standpipe, which should be as long as practical. Once the float valve closes, runoff goes into the tank. A weep hole

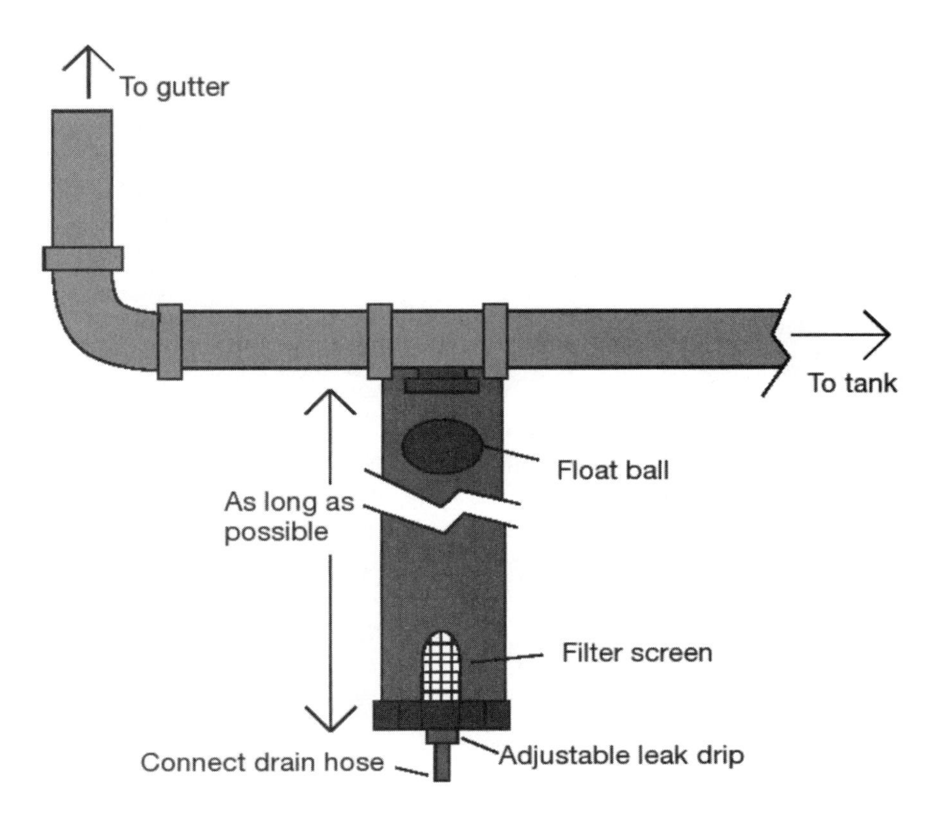

Figure 3-21. *A first-flush diverter with float valve. The screen requires periodic attention.* Oregon Dept. of Consumer and Business Services

or an adjustable valve drains the standpipe between rains. A cleanout fitting is provided at the base of the pipe.

Irrigation systems benefit from a roof washer, and potable water systems would hardly function without one. The device consists of one or two 30-micron (μ) filter cartridges in a fiberglass box. The washer goes on the inlet side of the system, usually at the junction between the guttering and the downspout that feeds the storage tank. The cartridge-type filters must be periodically changed.

Maintenance

Trim back branches that overhang the roof. Trees encourage animal life and, of course, shed leaves that clog screens and oxidize in the storage tank. Hose down the roof and clean the gutters, drains, and first-flush reservoir at least every six months. Repair any damage to the gutter leaf screens.

Rain, in its virgin state, is neutral. That is, it is neither acidic nor alkaline with a pH of 7.0. By-products of fossil-fuel combustion—sulfur dioxide from coal-burning generating plants and nitrogen oxides from automobile exhausts—lower the pH. It is not unusual to encounter rain with an acidic pH of 5.6 or less. Acidic water damages root systems and washes out nutrients essential for plant life.

Litmus paper or any of several inexpensive test kits detect the presence of acidity. Commercial meters provide pH numbers, but represent a fairly hefty investment. Electronics-proficient readers can build an equivalent meter around a Texas Instruments LMC6482 high-impedance operational amplifier connected to a pH probe and a digital multimeter. Probes, available from aquarium and scientific supply houses, cost around $50.

Raise the pH by adding a small, one- or two-ounce piece of limestone or an ounce of quick lime, soda ash, or hydrated lime to the tank. Give the material a week or so to dissolve and retest.

Two schools of thought exist about tank sediment, which contains a mix of benign and harmful microorganisms. The Cabell Brand Center, a well-respected environmental group, recommends that tank sediment not be cleaned if the system has a first-flush diverter. Other authorities are not so sanguine and recommend that the tank be cleaned at intervals ranging from six months to two years. In the Australian Outback, roving teams of specialists make their living by cleaning storage tanks.

You should also have the water tested periodically for E. coli. Only one strain of E. coli is harmful, but a high count can signal the presence of other, less benign bacteria. E. coli multiplies quickly, which makes testing for it inexpensive.

If you suspect contamination, sanitize the tank with unscented Clorox, which contains 5.25 percent of the active ingredient sodium hypochlorite. Mix one fluid ounce (30 ml) of the bleach with a quart of water per 500 gallons of tank capacity and pour it down the inlet pipe. A 1500-gal tank requires 3 oz. (90 ml), and a 2000 gal tank 4 oz. (120 ml). Do not exceed these limits.

Indoor Nonpotable

Many homeowners use rainwater for toilet flushing and clothes washing, activities that account for as much as 80 percent of indoor water consumption. A captive-air tank (see Chapter 5) will assure that the system has sufficient pressure.

While there are no hard rules about purity, the recommendations made to the Texas Legislature (Table 3-3) seem reasonable.

Table 3-3
Recommended Water Purity Standards for Single-Family, Indoor, Nonpotable Water Use (including clothes washing)

Contaminants	Limits
Total Coliforms	500 cfu/100 ml*
Fecal Coliforms	100 cfu/100 ml
Protozoan cysts	N/A
Viruses	N/A
Turbidity	< 1 NTU**
Testing interval	Annually

*cfu, or colony forming unit, measures the number of live micro-organisms in the sample
**NTU: Nephelometric Turbitiy Units.

Indoor rainwater plumbing should be clearly labeled as such. Otherwise, some future plumber might cross-connect these lines with potable water lines.

Potable Rainwater

Potable rainwater has not been a significant disease vector. Millions of people the world over drink it without ill effect. But that's little consolation to the victims of rainwater-induced Legionnaires' disease in Australia or the handful of Americans who have contracted Hantavirus Pulmonary Syndrome, which has a 40 percent mortality rate.

San Diego has made potable rainwater collection illegal. Other U.S. authorities recommend that the water you drink, use for cooking, and bathe in should be as toxin-free as current technology permits. However, these are merely recommendations. Even if they wished to do so, public health authorities do not have the resources to police millions of private water systems. EPA clean water standards apply only to municipal water treatment plants. The responsibility for privately sourced water rests with the owner.

Pollutants that are cause for concern include:

- **Coliform bacteria**—rod-shaped bacteria sourced from the feces of infected birds and rodents, and deposited on the rooftop collector. E. coli 0157H7 is the most infamous.
- Protozoan cysts—most often associated with diarrhea, but a variety of other diseases including typhoid. Giardia, and Cryptosporidium parvum are remarkably difficult to kill.
- **Viruses**—rare, but sometimes lethal. Hepatitis A is one of the more common virus-caused diseases.
- **Inorganic materials**—metals such as lead, copper, zinc, and aluminum stripped from roof coverings and gutters by acid rain. The EPA limits copper concentrations in public water supplies to 1.3 milligrams per liter (mg/L) and lead to 0.015 mg/L.
- **Organic chemicals**—numerous airborne contaminates associated with industrial activities.

- **Herbicides**—windblown herbicides, including 2,4-D (EPA limit 0.10 mg/L) and Atrazine (EPA limit 0.003 mg/L) enter rainwater as windblown aerosols.
- **Algae**

That said, rainwater is generally purer than surface water or water drawn from shallow wells. Water from these sources, which are subject to industrial and agricultural runoff, cannot be made safe to drink with the resources homeowners have at their disposal. Assuming that the aquifer has not been compromised, water from deep wells is purer than collected rainwater. However, purity is not something one can count on, and the water should be tested and treated as recommended by your local health authority. Filters and treatment options are the same as for RWH.

Objectives

At a minimum, one should aim at entirely eliminating fecal coliform, total coliform, protozoa cysts, and viruses. Turbidity should test less that 1.0 NTU (Nephelometric Turbidity Units). The solids that cloud water act as a shield for bacteria and viruses, protecting them from chlorination and UV light. Some state authorities suggest testing collected water for these impurities every three months; others say twice a year is sufficient. The City of Atlanta would have reclaimed rainwater tested annually for lead, cadmium, arsenic, and other heavy metals.

ANSI/NSF Standards

All components of the system—tank, filters, and treatment units must have ANSI/NSF food-grade certification. Polyethylene tanks come under ANSI/NSF Standard 61 and/or U.S. FDA regulation 21CFR 177.1520 (1) 3.1 and 3.2 for resins. Cartridge and carbon filters come under ANSI/NFS 51; UV light units under ANSI/NSF Class A. PVC plumbing products intended for potable water applications are nearly always marked NSF-PW or NSF-61. Food-safe copper tubing carries the same NSF-PW or NSF-61 designation, or else will be labeled NSF-DWV. When in doubt, consult your supplier.

Collection Surface

Most contaminants enter by way of the rooftop collector. Wash down the roof frequently, remove any overhanging limbs, and incorporate all the equipment options listed for irrigation water collection, that is, install a first-flush diverter and specify stable, nondegradable roofing materials (Table 3-4). Gutter screens are mandatory, and a roof washer is nearly so (Fig. 3-22). If a sealant or paint is applied, make sure the product carries an NSF food-grade certification. Aluminum (and not galvanized steel) is appropriate for flashing and gutters. Of course, no soldered connections should be used anywhere in the system.

Acid rain leaches roofing materials and, almost without exception, all other wetted surfaces. Unfortunately, the presence of a UV unit limits treatment options. Calcium carbon-

Table 3-4
Roofing Material Compatibility with Potable Water Systems

Material	Appropriateness
Aluminum-clad galvanized steel	Recommended. Most catchment-system contractors prefer this material. Some aluminum leaches into the water, but the health risk is not considered significant.
Slate	Recommended.
Clay / concrete tile	Recommended when treated with an NSF-approved food-grade sealant.
Tile	Recommended when treated with an NSF-approved food-grade sealant.
Asbestos shingle	Not recommended.
Lead, including lead flashing	Not recommended.
Galvanized steel	Not recommended.
Copper, including flashing	Not recommended.
Composite and asphalt shingle	Not recommended.
Treated wood or wood shingles	Not recommended.
Asphalt or composite shingle	Not recommended.
Tar and gravel	Not recommended.
Green roofs with vegetation	Open to question.

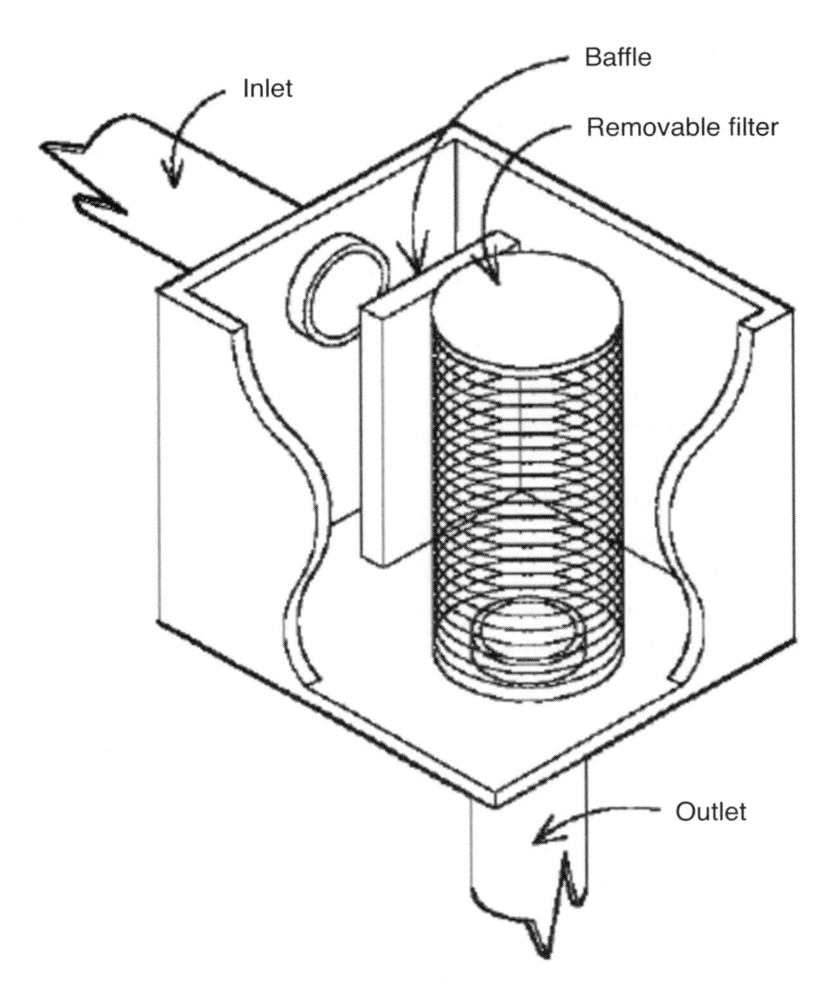

Figure 3-22. *What the industry terms a "roof washer" consists of one or two extremely fine filters that remove much of the organic matter that otherwise would collect in the tank.* Texas Water Development Board

ate, calcium oxide, or sodium carbonate pellets can only be introduced downstream of the unit. With expert plumbing, potable water from the municipal supply can be blended with rainwater to reduce acidity. A backflow preventer eliminates the danger of rainwater entering city mains. Even so, if you live downwind of an industrial area, you might entertain second thoughts about potable rainwater.

Filtration

The technology of filtration is complex with a vocabulary keyed to mesh size in micrometers. One micrometer (μm) is 1,000,000,000th of a meter, or 0.000039 inch. The period at the end of this sentence is about 500 μm in diameter. In terms of their construction, filters come in three varieties:

1. Particle filters, a category that includes bag and replaceable cartridge type filters.
2. Membrane filters that remove particles as small as 0.0001 μm.
3. Granular Activated Carbon (GAC) filters that work by absorption.

Table 3-5 provides a rough summary of the effectiveness of various filters.

Application

Begin by assuring that the stored water receives an adequate supply of oxygen: bacteria in tank sediment convert organic matter into carbon dioxide. The same bacterial action enables brooks and streams to do their own housekeeping. However, an overload of leaf particles and other organics create oxygen-starved anaerobic zones. The "good" bacteria die and harmful strains multiply. An ANSI/NSF 100-μm screen at the tank inlet removes much of this material. You might also install an electric or wind-powered fan at the tank vent and, if possible, arrange for three or four overflows a year—even if it means flooding the tank with municipal water—to drain off the oxygen-shielding oil film that collects on the water surface.

The inlet hose to the submersible pump should have a filtered and floating intake, as described for irrigation water. But, while irrigation water needs only coarse filtration, potable water requires a 5-μm filter. Both the fine-mesh inlet screen and the pump filter will need frequent attention.

The intensity of downstream filtration should reflect practices common to your area. If there is such a thing as a typical installation, it would consist of the two tank stages of filtration described in the previous paragraph, plus a 1-μm in-house filter, and some arrangement for disinfection.

Table 3-5
Filter Effectiveness

Filter	Pore Size	Pathogens Removed	Comments
Bag and cartridge particle	Variable	Protozoa	Inexpensive.
Microfiltration	10 to 1.0 μm	E. coli, Giardia cysts	This and finer-pore membrane filters clog easily and require regular replacement.
Ultrafiltration	0.1 to 0.01 μm	Some viruses	Same as above.
Nanofiltration	0.001 to 0.001 μm	Nearly all viruses, most naturally occurring organic matter and divergent ions that harden water.	Same as above.
Reverse osmosis	0.0001 μm	Turbidity, nitrates, most microbes and viruses, sulphate, lead, radium, and small amounts of some pesticides. Desalinizes.	Requires pre-filtration and a pump to restore pressure downstream of RO. Hard water damages the expensive-to-replace membrane.
GAC	Variable	When properly designed, will absorb small amounts of heavy metals, certain pesticides, and other organic compounds. Improves water taste and deodorizes.	Must be replaced regularly. Can become a haven for bacteria. Installed last in the system at the point of use.

Filtration of 1-μm or Larger Requires a Disinfection Stage

RO (reverse osmosis) filters down to the molecular level. However, RO units are expensive, power-consuming, pass only about a third of the water they process, and pose the problem of waste disposal in rural areas. If connected to an urban sewer network, the waste goes down the drain. You may also need to restore water pressure downstream of the RO unit.

Disinfection

Most RWH systems disinfect with one or more UV light units at the kitchen sink and other points of use. Some homeowners opt for chlorination, with the equipment located upstream of the pressure tank. Done right, chlorination requires a precision injector pump and settling tank large enough to provide two or more hours of contact time. Dosage depends upon the amount of chlorine injected and how long it remains in undisturbed contact with the water. Chlorine is a highly reactive oxidant, attacking plumbing and combining with organic material in water to produce trace amounts of carcinogens.

Ozonation is another option, used extensively in Europe and by some, perhaps more perceptive, American homeowners. Ozone destroys pathogens at least as effectively as chlorine or UV, but unlike chlorine leaves no harmful substances behind.

As Table 3-6 suggests, no treatment technology achieves 100 percent effectiveness. Some pathogens always get through. But well-designed and maintained RWH systems can provide better quality water than can be had from municipal systems, nearly all of which rely on chlorination and few of which conform with any regularity to Clean Water Act standards. Bottled water, which does not fall under EPA rules, is even more problematic.

Air Conditioner Condensate Recovery

With a small modification to its Science and Ecosystem Support Division laboratory cooling system, the EPA reduced freshwater consumption by a million gallons a year. The air con-

Table 3-6
Disinfection Options

Type	Pathogens Neutralized	Comments
UV	Bacteria and viruses	Requires a 5-μm or smaller prefilter, slow flow through the unit, and strict attention to maintenance. UV lamps dim with age and should be replaced every 10,000 hours.
Chlorination	Bacteria and viruses	A chlorine residue remains in the plumbing to increase contact time and effectiveness. The dosing pump, which should be of the positive-displacement type, requires careful monitoring. Ineffective against Cryptosporidium. Generates trace amounts of carcinogenic compounds.
Ozonation	Most micro-organisms and some pesticides	Expensive and requires prefiltration to remove solids that shield pathogens. Capabilities vary widely between brands and models, but none are effective against Cryptosporidium.

ditioning (AC) system consists of three roof-mounted air handlers, or evaporators, and a cooling tower (Fig. 3-23). Condensate from the air handler drip pans goes to the cooling tower where it replaces water lost to evaporation (Fig. 3-24).

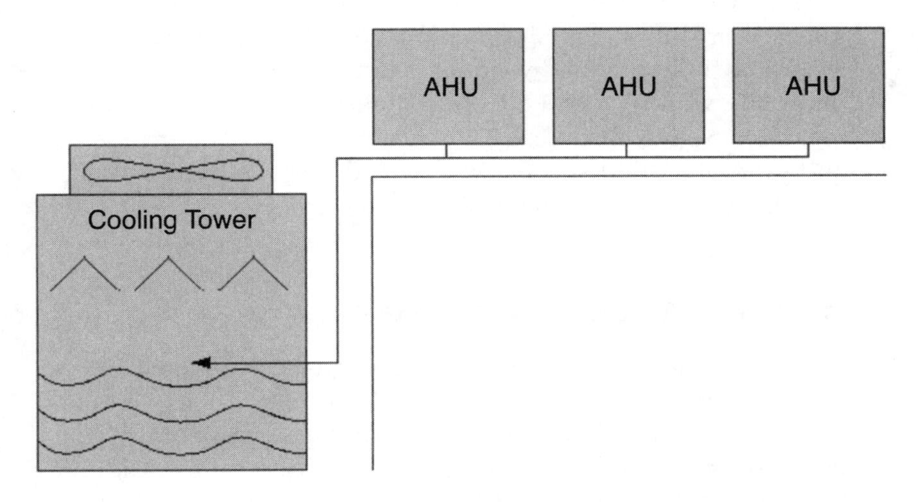

Figure 3-23. *Three air handler units provide replacement water for the cooling tower at the EPA's SEED laboratory.* EPA

Figure 3-24. *The smaller pipe picks up condensate from the air handler drain pans. Household AC units have a drain line under the pan.* EPA

Homeowners cannot expect anything near a million gallons a year. But a central AC unit can easily produce 10 gallons a day in hot, humid weather. Simply replace the existing drain line with PVC or PEX (cross-linked polyethylene) tubing. Because it is flexible, PEX is easier to work with in attics and other crowded spaces. We spliced the condensate line into the rainwater tank inlet pipe (Fig. 3-25).

A Word about Greywater

State authorities almost universally view greywater—the water from washing machines, dishwashers, and kitchen and lavatory sinks—as a mild form of sewage. Permits, on-site inspections and septic tanks are required. When used for irrigation, the water must be pumped from the septic tank and deliv-

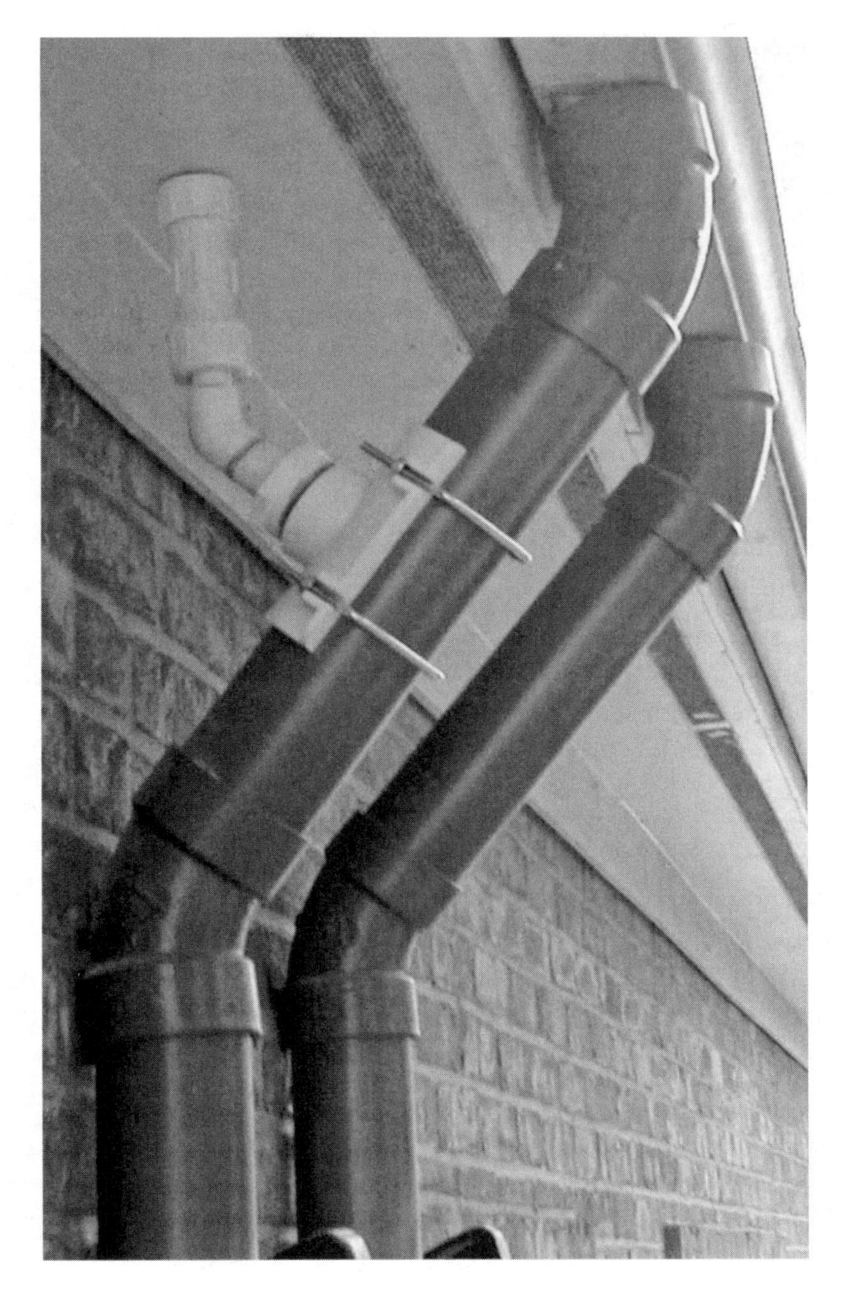

Figure 3-25. *Condensate from the attic-mounted AC unit goes into the rainwater catchment tank.* Tony Shelby

ered via underground perforated pipes, guaranteed to plug up within a year or so.

That said, it seems wasteful not to recycle the 40 gallons per load that conventional washing machines send down the sewer. One merely runs the drain hose to a plastic food barrel or other holding tank that has been fitted with a hose bib. Be careful to keep the drain hose outlet above the tank water level to prevent back-siphoning. The City of Tucson makes things easy by mandating that new construction have the necessary plumbing installed.

Apply the water to the base of ornamental plants and trees. The EPA warns against using the runoff on food crops. Generally the slightly acidic water is beneficial to plant life. Cottonwood, honeysuckle, burning bush, rabbit brush, junipers and most desert plants thrive on the stuff. Begonias, primroses, ferns, crape myrtle, azaleas, violets, and magnolias do not fare so well.

Use "natural" detergents (the kind without whiteners), enzymes, boron, borax, chlorine, and sodium. Apply the water to the garden within a day or so of collection. If you let greywater stand, it oxidizes into gunk. And, finally, do not recycle the E. coli-contaminated runoff from washing diapers or soiled bed linen.

Other applications seem less promising. Water from kitchen sinks and garbage disposals contains animal proteins, fats, and greases that the landscape could do without. Recycling bathroom water for toilet flushing has more possibilities, but shower and lavatory drains will not begin to keep pace with the demand for flush water. And the plumbing for these hybrid systems gets complicated.

However, some situations and living arrangements make greywater harvesting practical. Art Ludwig's beautifully illustrated *Create an Oasis with Greywater* and *Builder's Greywater Guide.,* both published by Oasis Design, are must reading for anyone who wishes to pursue the subject.

4

Wells

Groundwater collects in aquifers that, in North America usually consist of layers of sand, gravel, or silt deposited over an impermeable base material, such as clay or bedrock. Less common are aquifers formed by highly fractured granite, basalt, or limestone.

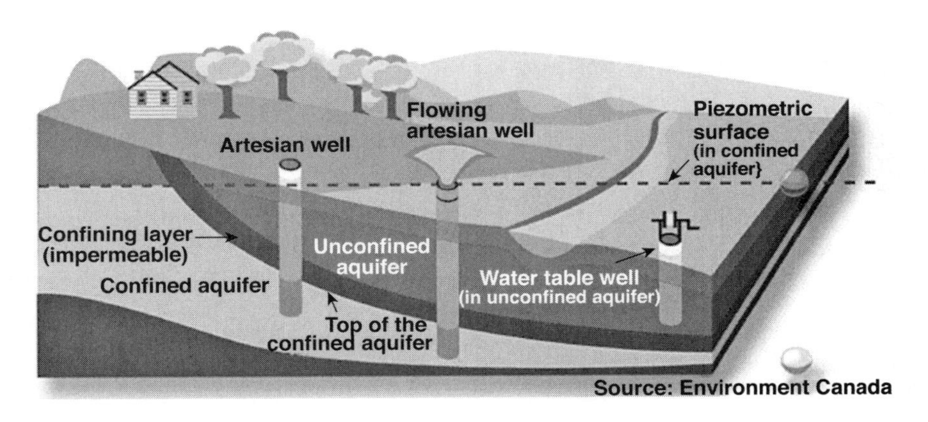

Figure 4-1. *Unconfined aquifers establish the water table and, as such, are candidates for shallow DIY wells. Confined aquifers are roofed by an impermeable layer of clay or rock. If a confined aquifer is under pressure, tapping it results in an artesian well.*

Well Basics

A water well consists of:

- **Borehole**—the wellbore (Figs. 4-2 and 4-3).
- **Casing**—PVC or steel pipe, typically 3½ in. to 6 in. in diameter, that lines the borehole. The casing extends down to a proscribed depth and stands at least a foot above the surface. It prevents loose soil, rocks, and sediment from entering the well and can also house a submersible pump or components associated with surface-mounted jet or hand pumps (see Chapters 5 and 6).

  The type of casing depends upon the drilling method. Most DIY drillers opt for PVC casing, since it costs only 30 percent as much as steel and is much easier to handle. Steel casing has the advantage of strength and, when

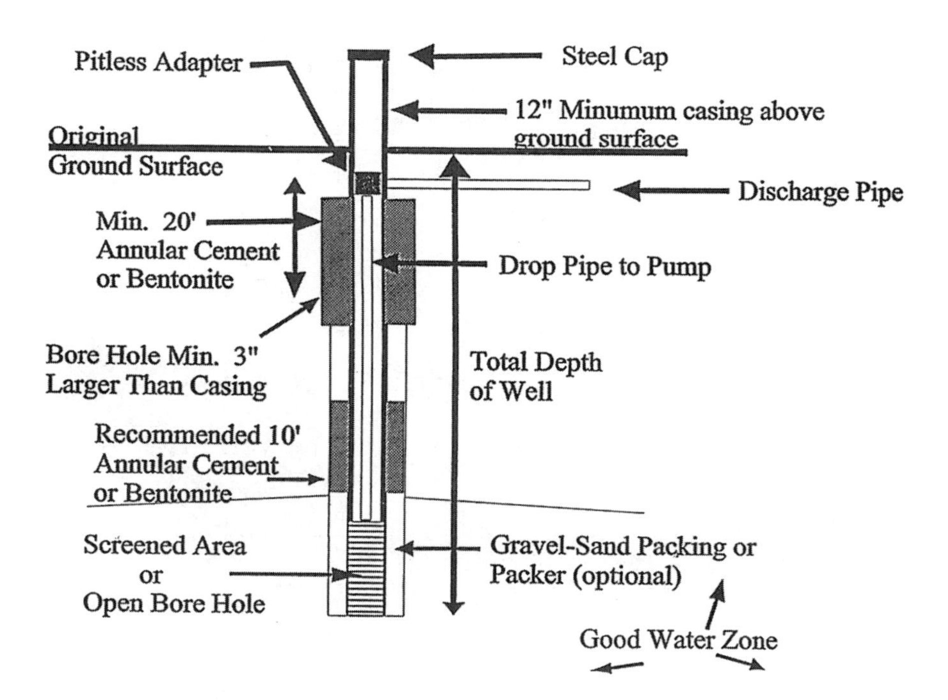

Figure 4-2. *Wells that tap into sand or gravel aquifers terminate in a screen that may be supplemented with a gravel pack.* State of Texas

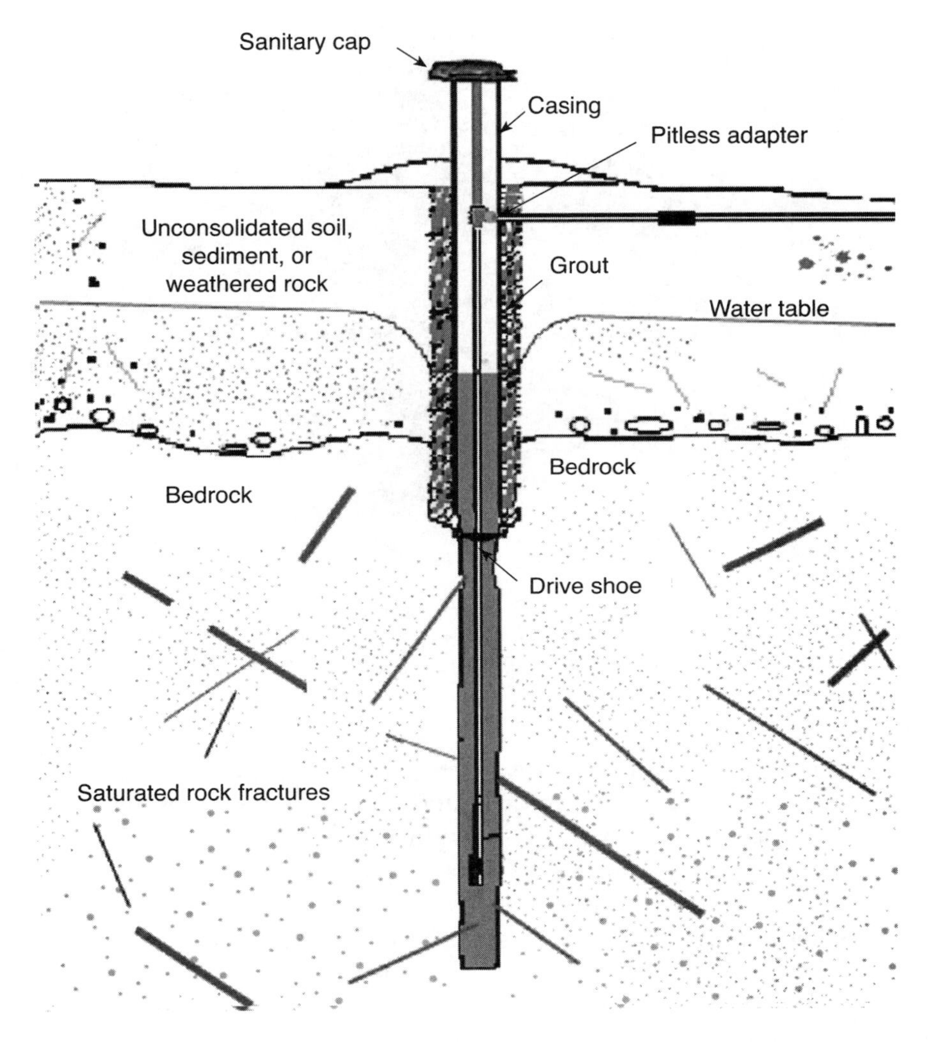

Figure 4-3. *Wells that draw from fractured bedrock do not require a screen.* Pennsylvania Department of Conservation and Natural Resources

fitted with a welded-on drive shoe, can be driven into the wellbore. Some professionals use nothing else.

- **Sanitary cap**—a vented and removable cap that closes off the top of the casing.
- **Pitless adapter**—used when the pump is offset from the wellbore (Fig. 4-4). The adapter functions as the

Figure 4-4. *A pitless adapter is a 90° ell that seals off the top of the casing and diverts flow to the surface piping.*
Pennsylvania Department of Conservation and Natural Resources

connection point between the vertical drop pipe and the horizontal discharge pipe. Standard practice is to mount the adapter a few feet (but within arm-reach) of the surface. The discharge pipe runs underground to protect it from impact and freeze damage.

- **Drop pipe**—also known as the riser, the drop pipe carries water from the aquifer to the surface.
- **Check valve**—a one-way valve that prevents water from draining back out of the drop pipe when the pump is idle. If the check valve mounts at the base of the pipe, it is called a "foot valve."
- **Gravel pack**—a gravel "envelope" surrounding the screen that functions primarily as a filter. Used in soft, unconsolidated formations.
- **Grouting**—a sealing material forced into the annulus between the casing outside diameter (OD) and the wellbore to protect the aquifer. Its composition varies: in the past drillers favored bentonite, a montmorillonite clay

that swells 500 percent when wetted. Most drillers now use Portland cement, either neat or mixed with bentonite or sand.
- **Screen**—a crude filter used to minimize sand intake.

The Screen

The screen admits water to the riser-pipe, while blocking large sand particles. During the development phase, the screen passes fine sand, "loosening" the well to improve water flow. As the well ages, a porous cake forms around the screen that filters out smaller particles. How long the pump lasts depends, in great part, upon the effectiveness of the cake.

Standard screens are 36 in. long, but the shape and configuration of the orifices vary. Slotted screens are numbered in thousandths of an inch: a No. 12 point has 0.012-in.-wide slots, a No. 14 has 0.014-in.-wide slots, and so on (Table 4-1). Mesh or gauze screens range from No. 40 to 90, with the higher numbers meaning finer filtration. Johnson and their imitators wrap the pipe with V-shaped wire, so that the slots for water passage widen, as shown in Fig. 4-5. This design,

Table 4-1
Screen size depends upon formation type

Formation	Slot size	Openings (in.)	Openings (mm)	Gauze size
Clay, silt	—	0.003	0.10	—
Fine sand	6	0.006	0.15	90
	7	0.007	0.18	80
	8	0.008	0.20	70
	9	0.009	0.25	60
Medium sand	12	0.012	0.30	50
	15	0.015	0.40	—
	18	0.018	0.45	40
	20	0.020	0.50	—
Coarse sand	25	0.025	0.65	30
	35	0.035	0.90	20
	50	0.050	1.27	—
Fine gravel	90	0.090	2.30	—

Source: Michigan Department. of Natural Resources & Environment

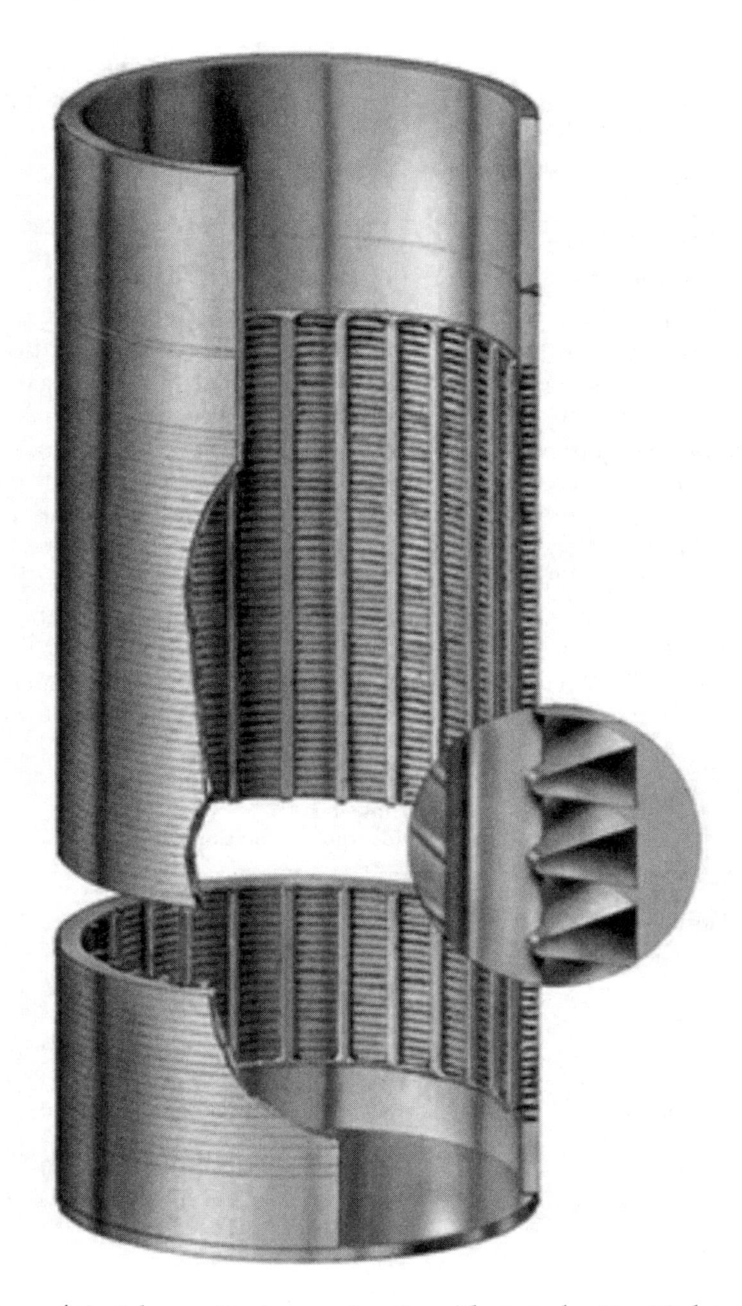

Figure 4-5. *Johnson V-wire construction. The popular No. 10 slot configuration blocks particles larger than 0.010 in.*
Michigan Resource Management Division

upon which the company was built, resists clogging. Screens can be mounted in tandem to improve the flow from thick aquifers.

Note: *Doubling the well diameter increases yield by 10 percent; doubling the screen length exposed to the aquifer increases yield 100 percent.*

Buying Property with a Well

Home mortgage companies usually insist on a water test, but the condition of the well and associated plumbing is equally important. Most states require drillers to submit a well log, or driller's report, to the appropriate agency. The well owner also receives a copy. Some well logs merely provide the reference number of the well, the completion date, and the yield; others go into great detail, describing formations encountered, equipment types and model numbers, and the drilling method. This information can be quite useful in the event that something goes wrong.

The salient consideration is the age of the well. Table 5-2 in the following chapter lists what the EPA has determined to be the normal life spans of wells and associated equipment. It makes for somber reading.

The owner should also be able to supply maintenance and inspection records, together with the results of water tests. The county health department should have water tests on file.

In addition to written records, make a careful inspection of the well to determine that it is at least 100 feet from and preferably uphill of septic tanks, animal pens, and other sources of contamination. All things being equal, the deeper the well, the safer the water.

The casing should stand a foot or more above the ground to be safe from flooding. The well cap should be tight with the vent pointed down and screened. Surface plumbing should have backflow protection.

Hiring a Driller

Where and how much water will be found are questions that can only be answered by drilling. Some homeowners put their faith in dowsing; others hire well surveyors, many of whom make impressive scientific claims for their work. But nobody offers a guaranteed outcome. Consequently, the well should be drilled before building on the property. People have built fine houses only to discover that they must survive on trucked-in water.

The proposed site must be accessible to drilling equipment, which means that it should be clear of underground and overhead power lines and overhanging trees—some rigs have 50-ft booms—and situated on reasonably flat ground. If in doubt about the location of subterranean power or gas lines, contact the utility.

Well permits are usually handled at the county level. There will be fees to pay and an inspector sent out to verify that the well site conforms to health and environmental regulations (Fig. 4-6). Most of these regulations establish setbacks from known sources of contamination.

Concern about protecting well-water quality has deep historical roots. Sir Thomas Gates, the second governor of the Jamestown settlement, was very much aware of the danger contaminated water poses. His proclamation of 1610 spelled out what precautions were to be taken:

> There shall be no man or woman dare to wash any unclean linen, wash clothes,... nor rinse or make clean any kettle, pot or pan, or any suchlike vessel within 20 feet of the old well or new pump. Nor shall anyone foresaid within less than quarter mile of the fort, dare to do the necessities of nature, since by these unmanly, slothful, and loathsome immodesties, the whole fort may be choked and poisoned.

Sir Thomas only had to deal with what we now call blackwater and greywater. Those pollutants are still with us, plus hundreds of others from chemically enhanced agriculture and industrial processes. But the basic strategy of separating the well as far as possible from known sources of contamination remains unchanged. The earth is our most reliable filtering

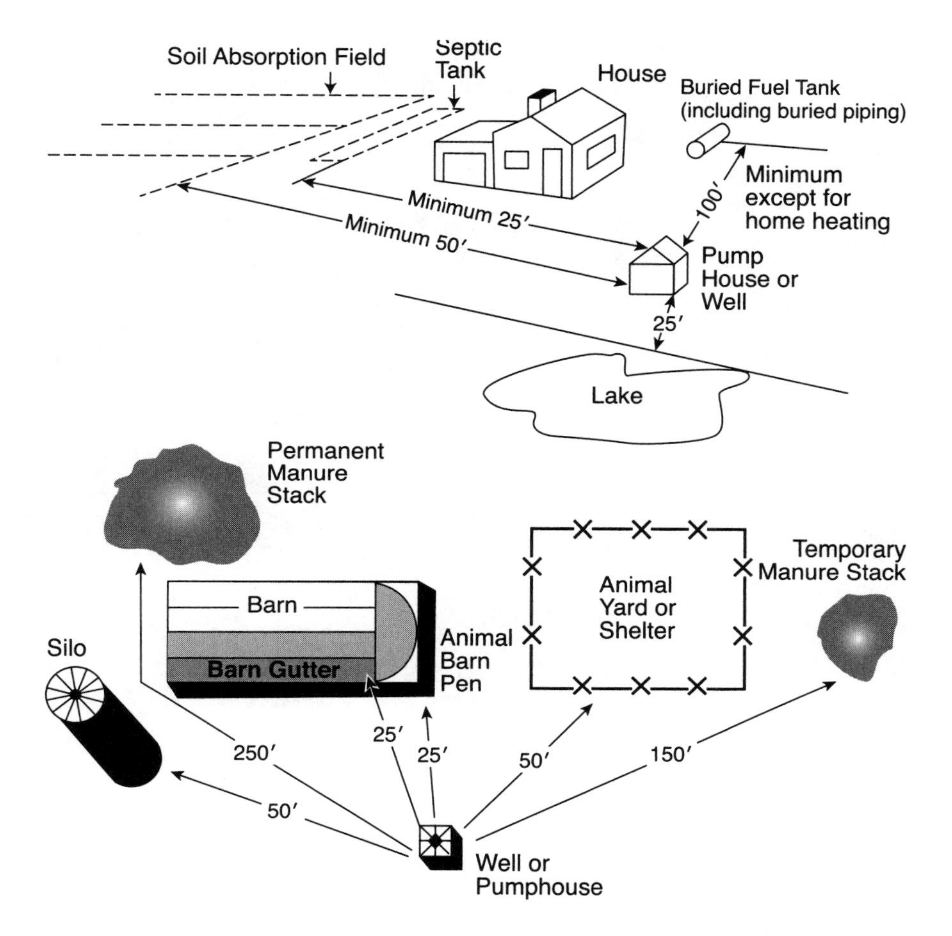

Figure 4-6. *Well locations approved by the Wisconsin Department of Natural Resources. Rules in other states vary in detail, but all agree that water wells should be outdoors (not in basements), remote from known sources of contamination, and protected from flooding.*

mechanism. In addition, the wellbore must be securely capped and the interface between the casing and the earth sealed with grout or cement.

Most professional drillers use a rotary rig such as the one shown in Figure 4-7. These rigs exhibit good rates of penetration, but require large quantities of drilling fluid ("mud")

Figure 4-7. *Rotary drilling rig. Drilling fluid recirculates through the mud pit shown in the foreground.*
Michigan Resource Management Division

to circulate cuttings out of the hole. Bentonite-based mud compromises the permeability of the aquifer to some degree. The older method, favored in some parts of the country, is percussion drilling. A cable-tool rig, such as the much-prized Bucyrus-Erie 22W, raises and drops a weighted, chisel-like bit that hammers its way down. The Bucyrus weighs 9750 lb, carries a 40-ft mask, and can drill to depths of 1000 ft. No drilling fluid other than a small amount of water is needed. Because it proceeds by fracturing, percussion drilling improves the flow from hard-rock aquifers. A cable-tool driller, having long since amortized his rig, may give you a slightly better rate than a rotary driller. It's worth inquiring.

The drilling contractor should have:

- **Certification.** Ask to see the contractor's state certification and note whether it is up to date. Some states have multiple levels of certification that give recognition to the more experienced drillers. Membership in professional societies may indicate something about pride in work and openness to new technology.
- **Insurance.** The contractor should be insured and, in some states, bonded. Ask to see proof.
- **Work history.** Follow up on references and, if possible, do so in person.
- **Professional knowledge.** Some estimate of the contractor's expertise can be had by inquiring about the nature of local geology and problems encountered. The contractor should paint you a coherent picture and, in the process, reveal something about the philosophy that guides his or her work.

Try to arrange a visit to the shops on your short list. Spend some time "nosing around" to get a feel for the place, a sense of how things go. The equipment, which need not be new, should be put away clean, without oil leaks and rust.

Do not be guided by price alone. There are dozens of ways, ways that you will not discover until years later, to skimp on quality. Likewise, avoid fixed-cost, turnkey jobs. The contractor needs some freedom to maneuver.

When possible, have all work done by a single firm. If responsibility splits between drilling, plumbing, and electrical contractors, any problems that arise are always someone else's fault.

And, finally, read the contract carefully to understand exactly what you're paying for.

DIY Drilling

Do-it-yourself drilling is a form of gambling, with stakes increasing as you go deeper. Sometimes you have to fold and leave the table. But with ingenuity and persistence, the odds are that you'll strike water, the most precious commodity on earth. That explains why DIY drillers hardly ever stop with a single well. The mix of sweat, uncertainty, and exaltation is too strong to resist.

In order to drill a well without elaborate machinery, realize the following:

- The water table should be shallow, preferably no more than 20 feet or so below the surface. Deeper auger-drilled wells are possible, but the workload increases with depth. At some point, the balance tilts in favor of calling in a professional with the proper equipment.
- Not much by way of yield should be expected. A 20-ft well does fine if it delivers 5 or 6 gallons per minute (gpm), which is the amount of water that a single hose bib flows. And unless one takes really heroic measures shallow groundwater cannot be made safe to drink.
- Formations must be soft and unconsolidated. Hard clay, silt, gumbo, and fine sand are difficult to drill through. Boulders, cobble, and bedrock are impossible to penetrate with hand-drilling methods.

Homework

The British Army saying that time lost in reconnaissance is never wasted applies in spades to DIY well projects. Check with your state authorities, beginning with the county agent, to learn as much as possible about the potential for extracting groundwater in your area. If the aquifers lay close to the surface and the formations can be breached by muscle power, determine what legal restrictions apply. Some states rule out DIY wells entirely and others insist that aspects of the operation, such as installing an electric pump, be performed by a licensed professional.

You will need to know the following:

- Depth of the water table
- Nature and hardness of formations that will be encountered
- Type of aquifer and its yield potential

- Preferred method of DIY drilling
- Advice on well screen type and slot sizes
- Whether casing is required, and if the casing must be cemented or grouted
- Type of well seal
- Suggested decontamination procedures

Talk with local well owners to learn about any problems they may have encountered. And last, but surely not least, contact the U.S. Geological Survey for information about drilling in your region.

Percussion Drilling

The simplest way to tap underground water is to hammer a pipe into the ground. Of course, the soil must be relatively easy to penetrate.

Most percussion, or driven-point, wells are quite shallow, rarely more than 25 feet, although deeper wells have been completed in very soft formations. Yields are small, no more than 10 gpm.

The Drive Pipe

The drive pipe usually doubles as the drop pipe. In other words, the same pipe used to drive the well home supplies water to the surface. A drive pipe consists of 5-ft or 6-ft sections of 1¼-in. or 2-in. galvanized pipe. Smaller diameter 1¼-in. pipe is for very shallow wells served by hand or electric surface pumps.

Deeper wells call for 2-in. drop pipe to accommodate a downhole pump or that part of a surface pump that actually moves water. However, an EMAS-style (Escuela Móvil de Agua y Saneamiento or Mobile School for Water and Sanitation) hand pump that fits inside a 2-in. pipe may be the best choice. See Chapter 6 for instructions on how to fabricate one of these remarkable devices.

Some state authorities mandate that wells be cased, regardless of depth. Pumps for these wells feed through a drop pipe concentric with the casing (Fig. 4-8).

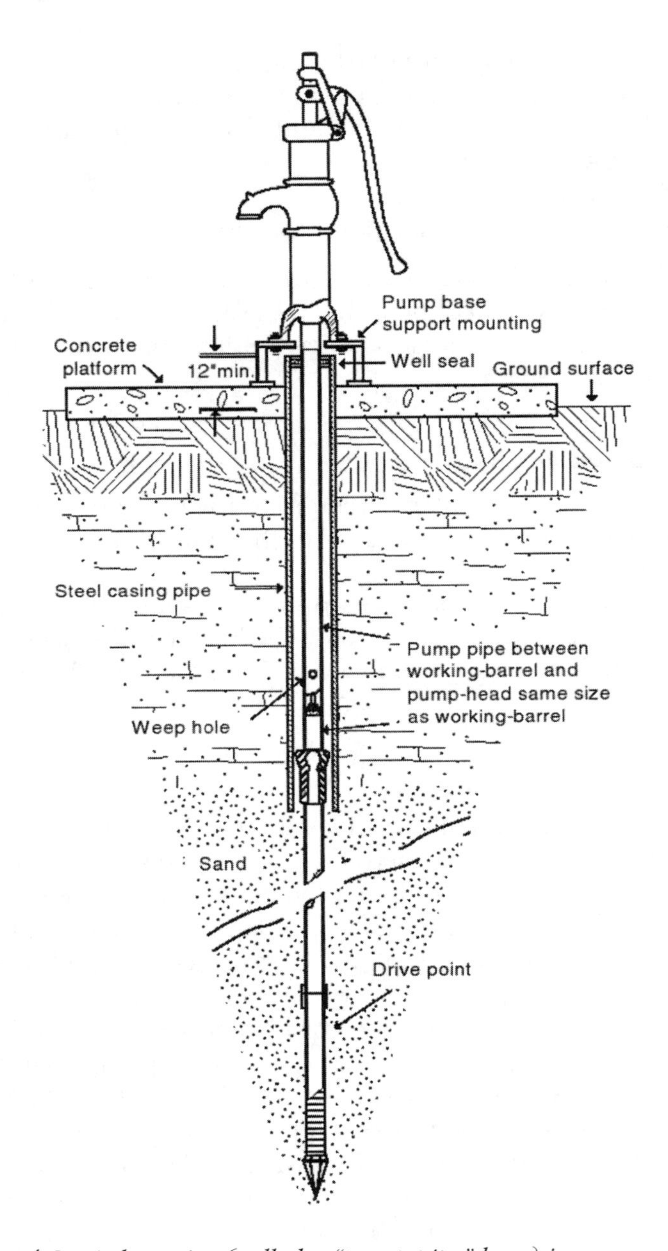

Figure 4-8. *A drop pipe (called a "pump pipe" here) is recommended for wells that yield drinking water. If the casing were used to deliver water, it would be subject to a vacuum during pumping. The weep hole prevents freezing.*
Wisconsin Department of Natural Resources

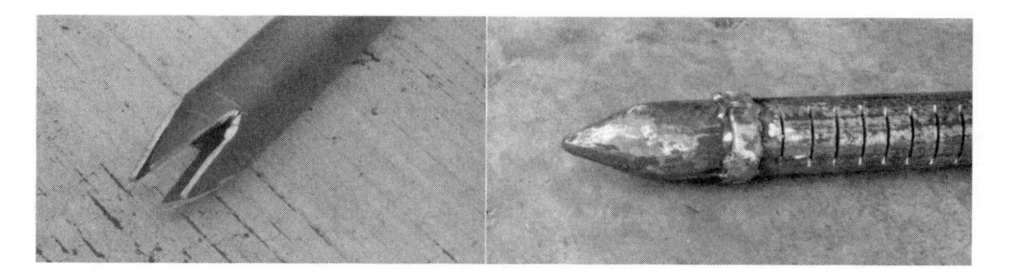

Figure 4-9. *One way to make a drive point: Cut four triangular slices from the pipe as shown on the left. Form into a cone, weld the seams, and cut the screen slots. The ring, while difficult to see in the photo, reams the hole slightly oversized to help prevent sticking.*

Drive Points

A drive point, also known as a sand, well, or "bangable" point, consists of a point and a screen assembly that admits water to the drop-pipe ID, while blocking large sand particles. Drive-point screens can cost several hundred dollars, and some readers may want to fabricate their own (Figs. 4-9 and 4-10).

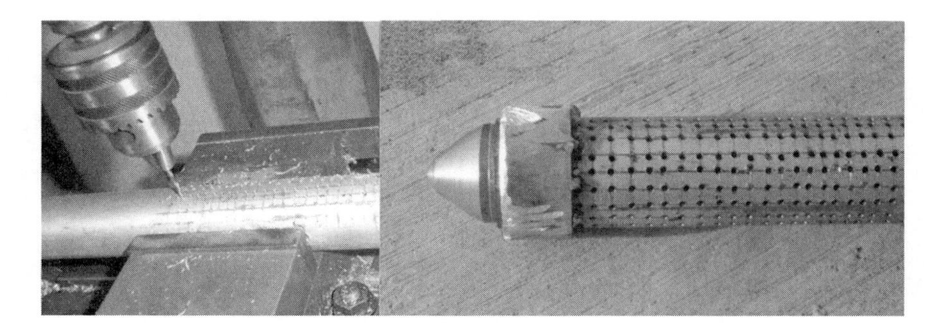

Figure 4-10. *Something like 1000 holes were drilled, but this technique gives control over orifice size without overly compromising the collapse strength of the drive pipe. The point was turned on a lathe.*

Drive Couplings

Unlike conventional pipe couplings, drive couplings butt the pipe ends together so that impact forces pass through the pipe and not the threads. Some drive couplings incorporate skirts that extend over the non-threaded portion of the pipe to give lateral support to the joint.

The alternative is to use standard Schedule 80 couplings. If hard going is expected, either cut the threads back so that the pipe sections butt against each other or weld the couplings to the pipe. You will need one less coupling than the number of 5-ft or 6-ft drive-pipe sections.

Caution: *Grind off the galvanized coating adjacent to the welds and work in a well-ventilated place. Zinc vapor is toxic.*

Drive Caps

A drive cap is a heavy-duty pipe cap that screws on the upper end of the drive pipe to absorb hammer impacts. As far as I know, only the Simer Pump Company makes alloy-steel drive caps; the others are frangible iron castings.

Drivers

Many wells have been driven by pounding on the drive cap with a sledge or a heavy wooden mall. When the pipe goes down straight, there is nothing wrong with this technique. But off-center blows send the pipe down at an angle and can stress the coupling threads enough to cause air leaks, discovered when the pump fails to lift water. You can rent the proper tool—a slide hammer—from a rural hardware store. The lighter versions center over the drive pipe and are hammered down; heavier models mount on a tripod and drive by their own weight. (Fig. 4-11)

Procedure

Step 1. Use a posthole digger or an auger to make the starting hole, which should be vertical and as deep as the tools permit. If the ground is hard, keep the hole filled with water for several days.

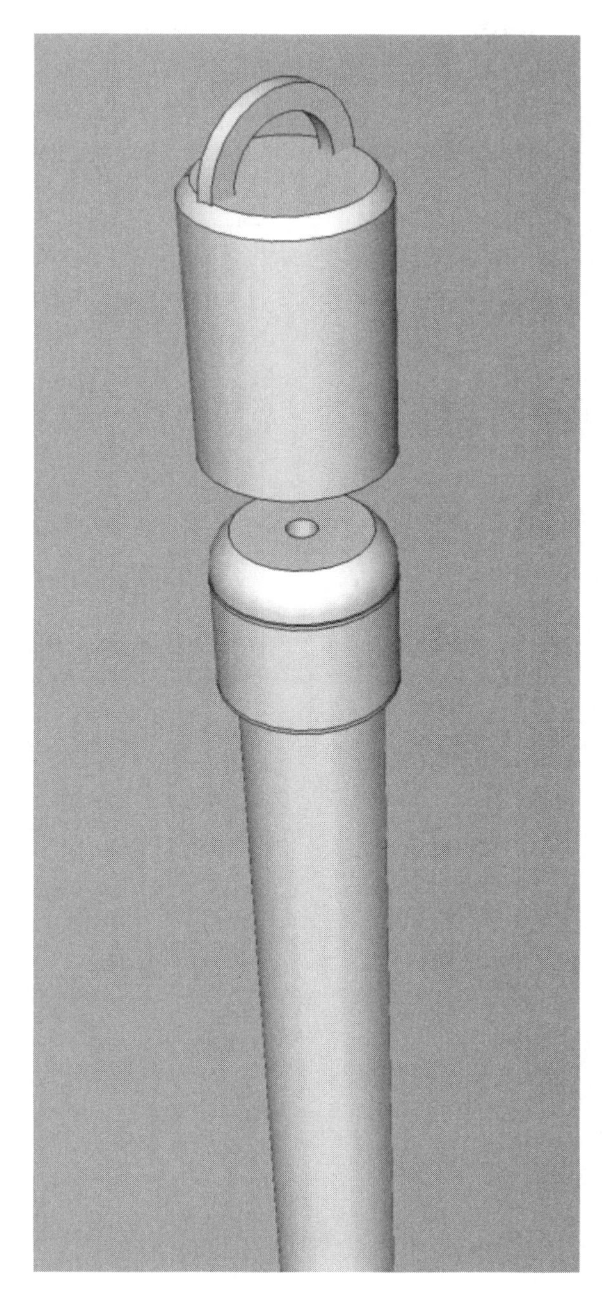

Figure 4-11. *A well driver for use with a tripod. The driver is lifted with a rope and dropped against the well cap.*

Step 2. Fill the openings in the screen with bar soap to prevent clogging. Soap also acts as a lubricant.

Step 3. Make up the first section of drive pipe to the drive point. Clean the threads, seal with Teflon tape, and tighten down hard with pipe wrenches.

Step 4. Thread the drive cap on the upper end of the pipe and wrench it down tight.

Step 5. Begin driving. Tighten the cap after each impact and make frequent checks with a level to assure that the pipe remains vertical. The way the drive pipe responds to impact gives clues about the nature of the formation (Table 4-2).

Should you encounter hard rock or a boulder, start over in another location. Fit a pipe clamp over the drive pipe and, with a pair of hydraulic jacks, lift it a few inches. Once freed, the pipe should come out easily.

Step 6. When the drive cap is about 6 in. above the ground, remove it, and make up the second section of drive pipe. Seal the coupling threads with Teflon tape, and tighten down hard.

Step 7. Repeat the above step, adding sections of pipe as the well progresses.

Table 4-2
The effects of formation on well driving

Formation Type	Progress	Rate of Descent	Sound of Hammer Blow	Rebound
Soft, wet clay	Easy	Rapid	Dull	None
Gravel	Easy	Irregular	Dull	None
Coarse sand	Easy	Irregular	Dull	None
Fine sand	Difficult	Variable	None	Frequent
Hard clay	Difficult	Slow and regular	None	Frequent
Bedrock and boulders	Nearly impossible	Little or none	Loud	Pronounced and may involve the entire drill string

Step 8. Periodically stop and fill the pipe with water. When the water drains out within a minute or two, the aquifer has been reached. You can also test for water with a weighted string or a pitcher pump like the one shown back in Fig. 4-8.

Step 9. Once you have reached the aquifer, go down several feet deeper to compensate for the drop in water level during pumping.

Step 10. The drive pipe should stand about a foot above ground when an electric surface pump is fitted. A pitcher pump mounts higher for ease of use.

Step 11. Fill the annulus between the riser and guide hole—the hole made with a posthole digger—with cement or a wet clay slurry.

Step 12. Purge the well. One method is to lower a garden hose to the bottom and pump until the water clears. You can also short-cycle a pitcher pump, moving the handle at less than full stroke so that its inlet check valve remains open. Oscillating a capped pipe at total depth has a more powerful effect. The rapid up-and-down motion sends reverse flow through the drill-point screen, opening clogged orifices.

Jetting

Jetting, or washing, is a less brutal way of accessing ground water. PVC tubing keeps costs low. But as with other DIY drilling techniques, the formation must be quite soft, the water table shallow, and expectations of production low. And one must already have a source of water to soften the formation and lift the cuttings.

Most wells are jetted with Schedule 40 PVC, although some practitioners prefer Schedule 80, which is more rigid and threaded. The head, made up from a 2-in. ell for use with a single garden hose, or a tee for two hoses, mounts to the PVC pipe with a length of radiator hose (Fig. 4-12). The bit consists of a 2-in. pipe coupling with teeth hacksawed on its lower end. The lower section of PVC pipe is slotted or drilled as shown back in Figs. 4-9 and 4-10 to serve as the screen.

Figure 4-12. *The wellhead assembly. Hose clamps permit the wellhead and handle to be quickly removed and fitted to another section of PVC pipe. The vent hole is circled for clarity. When water pours out tof he vent, lift the assembly a few inches to clear the clogged bit.* Photo by Wheeler Dial

You will also need to fabricate a handle that secures to the PVC pipe with radiator hose clamps.

Procedure

Step 1. Begin by cutting teeth with a handsaw in the end of a section of 2-in. PVC pipe (Fig. 4-13).

Step 2. Make up the head assembly. In some parts of the country, hardware stores sell jetting kits with all the necessary components. But it's simple enough to cobble together a drilling head from readily available PVC plumbing parts.

Note: *Use PVC primer on all joints.*

Figure 4-13. *The bit consists of a toothed PVC coupling.*
Photo by Wheeler Dial

Step 3. With a posthole digger, make a starting hole deep enough to clear turf and tree roots.

Step 4. Position the PVC pipe vertically in the starting hole, mount the handle at a convenient height on the pipe, and turn on the water.

Step 5. Commence drilling (or as drillers say, spud-in) by applying a moderate downward force on the pipe while, at the same time, rotating it back and forth (Fig. 4-14). The cuttings float up through the annulus and puddle at your feet. Progress through soft formations is quite rapid (Fig. 4-15).

Step 6. Make frequent checks with a level to verify that the pipe goes down straight.

Figure 4-14. *Spudding-in*. Photo by Wheeler Dial

Step 7. When the first section of pipe stands about a foot above ground level, disconnect the jetting head and add another section of PVC pipe.

Caution: *As you wait for the PVC cement to dry, periodically lift the pipe off the bottom and turn it. Otherwise the pipe will stick.*

Step 8. Circulation can be lost as water-bearing sands are penetrated. Should this happen, water no longer spills out of the annulus and progress stops. The best recourse is to remove the jetting head and complete the well with a drive point and screen as described in the previous section.

Figure 4-15. *Nikki made rapid progress, but had a bit of difficulty keeping the hole vertical.* Photo by Wheeler Dial

Step 9. Once the reservoir has been tapped, measure the depth of water inside the pipe with a weighted string. Continue jetting until the well is deep enough to wet the length of the screen plus an additional 24 inches to compensate for pump drawdown and seasonal variations in the water level.

Step 10. Lower a garden hose into the well to backflush the screen. An absence of return, that is, if flush water does not rise to the surface, is confirmation that the well opens to an aquifer.

Pump-Assisted Jetting

Flushing cuttings with a garden hose may not be possible in rural areas. A better approach is to recirculate the water with a portable pump and a settling tank, which can be no more elaborate than a mud pit lined with a plastic sheet (Fig. 4-16).

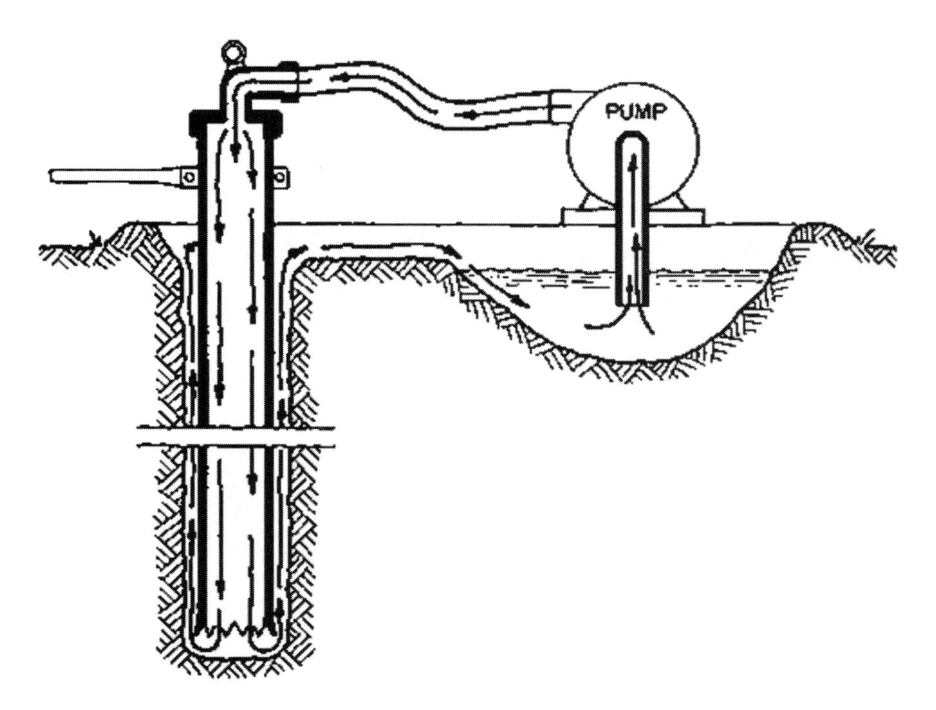

Figure 4-16. *This drawing illustrates pumping through the casing. The pump can also be connected to a drop pipe with returns going to the surface through the casing, rather than the annulus as shown here. A centrifugal pump delivers the high volume needed for rapid jetting and is less sensitive to solids than other types of pumps. The mud pit should be lined with plastic to conserve water.* UN Natural Resources and Environmental Development

You might also mix a little bentonite with the water to stabilize the borehole and give some protection against pipe sticking.

DIY Gasoline-Powered Pump

Most DIYers have a small utility engine lying around collecting dust. Hook the engine up to an automotive water pump, put wheels on it, and you will save hundreds of dollars over an equivalent commercial unit. A horizontal-shaft engine simplifies things, although a vertical-shaft rotary-mower engine will work, if the engine can run without a blade. Many vertical-shaft engines use the blade as a secondary flywheel.

Once you have a prime mover, the choices of a pump come down to three:

1. **A commercial pedestal pump.** The term "pedestal" means that the pump has a mounting stand and provision for a belt drive. Unfortunately pumps of this type cost $200 or more.
2. **A centrifugal pump intended to be driven by an electric motor.** These pumps bolt to the motor with the impeller supported by the motor shaft bearings. The lack of bearing support means that the pump must mount directly on the end of the crankshaft and dead in-line with it. Alternatively, you can drive the pump from a fabricated jackshaft, but in either case, the alignment must be precise. Mounting difficulties aside, pumps of this type are certainly worth consideration.
3. **An automotive water pump.** This pump has bearings that support the radial loads imposed by their belt drives, which simplifies mounting. Used pumps can be had for a few dollars. But choose carefully. Most bolt directly to the engine block and have no cover over the impeller. Look for a pump that does not have its impeller standing proud beyond the pump body casting. A flat cover, spaced off the pump body with a thick gasket, is easy enough to fabricate. The cover must also include an inlet fitting aligned with the impeller hub.

For want of anything better, we settled on a small-block Chevrolet V-8 pump. The pump draws from the large-diameter hose connection and discharges through the two runners extending out on each side of the pump body. We mounted the pump on a steel plate with ½-in. aluminum spacers (Figs. 4-17 and 4-18). One of the runners was blocked off and the other had its aluminum spacer threaded to accept a ¼-in. pipe. The somewhat awkward mounting position relative to the engine was dictated by the need to maintain the correct direction of pump rotation. We also had to modify a drive pulley (the original could not be found), which could have been smaller for an increase in pump rpm and efficiency. But the pump works well. The added water pressure and volume over a garden hose makes jet drilling go really fast.

Figure 4-17. *A Kohler 5-hp engine provides power for the Chevrolet V-8 pump.* Photo by Wheeler Dial

Figure 4-18. *A close-up view of the water pump.* Photo by Wheeler Dial

Drilling

DIY wells can also be drilled with an auger. This technique is more versatile than jet or percussion drilling in that it can be used for fairly deep wells. Auger drilling works best in soft, moist soils that adhere to the bit, which must be periodically brought to the surface and emptied. Dry drilling is the norm, but hard soils and loose sand can be wetted with moderate amounts of water. Deep wells require that the casing, usually PVC, be set almost continuously to prevent cave-ins and a drill-string sticking.

A two-bladed auger, similar to those used for making postholes, can move quite rapidly in very soft formations (Fig. 4-19). A helical auger is less sensitive to formation hardness and tends to drill straighter than the two-bladed type.

Figure 4-19. *Types of hand-operated auger bits.* U.S. Department of the Interior, Bureau of Reclamation

Figure 4-20 shows one of these wells being drilled in a sand dune in eastern Mexico. Four men rotated the string while another provided weight on the bit. Progress was slow. As the well deepened, the going got hard and the

Figure 4-20. *A drilling crew at work.*

bit twisted off. Days passed in fruitless attempts to retrieve it. A second hole had to be drilled. In all, more than two weeks were required to take the well to its target depth of 80 meters (260 ft)

Completing

In many jurisdictions, shallow DIY wells are unregulated, but the obligation exists to protect the groundwater source. The annulus must be sealed with cement or bentonite slurry packed down hard. The casing should be capped, as shown back in Figure 4-8, and the wellhead centered on a 2 ft^2 concrete slab at least 4 in. thick.

Most shallow wells are fitted with a hand-operated pitcher pump. Unless the pump has a check-valve positioned low enough in the drop pipe to be submerged, the pump will need to be primed before each use. A surface-mounted jet pump can sometimes prime itself without the help of a check valve, but don't count on it.

Shock Disinfection

The discussion that follows applies to wells deep and productive enough to be connected to household plumbing. As pointed out earlier, attempts to make water from shallow jetted-in or percussion-drilled wells safe for drinking are futile. Even if disinfection is successful, a heavy rain or a passing cow can re-infect the water table.

Shock disinfection is an emergency procedure, above and beyond the normal filtration and treatment arrangements necessary to make deep well or collected rainwater potable. For information about filters and treatment techniques, see the section on "Potable Rainwater" in Chapter 3.

State health departments mandate that newly completed, repaired, or flooded-over wells be disinfected with a 50 parts per million (ppm) solution of chlorine. Well owners who take this responsibility upon themselves rather than hiring a professional to do it generally use Clorox bleach (sodium hypochlorite), which contains 5.25 percent chlorine. Other brands have less of the active ingredient. Avoid bleaches containing perfumes or fabric softeners. Bleaches labeled as "safe for all fabrics" are by definition low on chlorine.

Warning: *Mixing household bleaches with ammonia or other chemical agents generates toxic gases.*

Procedure

Step 1. Disconnect the water supply to washing machines, water softeners, dishwashers, and other appliances that may be damaged by exposure to chlorine. These devices must be disinfected individually in accord with the manufacturers' guidelines.

Step 2. If an electric pump is fitted, turn off its power at the switch box. Power should not be restored until Step 6, and then only temporarily.

Step 3. Uncap the well.

Step 4. Using a plastic bucket, add one gallon of water to the amount of Clorox indicated in Table 4-3. Increase the amount of bleach proportionally for casing diameters greater than 6 in. and for wells deeper than 200 ft.

Warning: *Clorox and other commercial bleaches contain lye that damages eyes and skin upon contact. At a very minimum, wear eye and hand protection. If contact is made, wash the affected area immediately with water.*

Step 5. Using a plastic funnel, pour the solution into the well (Fig. 4-21).

Step 6. Pump the well out for at least ten minutes until the water runs clear. Drain the output through an outdoor faucet. The discolored water should not be permitted to enter household plumbing and must be disposed of properly. It is deadly on vegetation and on the bacteria in septic tanks.

Step 7. When you detect the smell of chlorine, recycle the flow back through the well as shown in Figure 4-22. Let the water run for about two hours.

Step 8. Stop pumping and seal the well.

Step 9. At this point, you have disposed of discolored water and recycled chlorinated water through the well. What remains is to disinfect the surface plumbing.

Table 4-3
Clorox requirements to achieve a 50-ppm concentration of chlorine in well water. These recommendations apply to the well only, without reference to surface piping or holding tanks.

Well Diameter (in.)	Well Depth (ft)		
	0 to 50	51 to 100	101 to 200
2 or less	1/8 cup	¼ cup	½ cup
3 to 4	½ cup	1 cup	2 cups
5 to 6	1 cup	2 cups	1 quart

Figure 4-21. *Adding chlorine to a well. Note that the lady wears goggles and rubber gloves.* Minnesota Department of Health, Well Management Section

Start the pump and turn on each fixture sequentially—the cold and hot water faucets, the showerheads, the washing machine, and the toilet—until the water smells of chlorine. That should not require more than a minute or so. Shut off the water and repeat the operation at the next fixture.

Figure 4-22. *Once water runs clear, divert the flow to the well.*
Minnesota Department of Health, Well Management Section

Chlorine test paper, available from swimming-pool suppliers, can also be used.

Step 10. With the pump off, allow the chlorinated water to remain in the system for 12 hours or more.

Warning: *Residents should not draw water (which may be available from the pressure tank even if the pump is shut off) during this period. The highly chlorinated water is dangerous.*

Step 11. To make the water safe for human use, it is necessary to remove all traces of chlorine. Start the pump and connect a hose to an outdoor bib. Let the water run until no trace of chlorine remains. This procedure can take as long as 24 hours for deep wells. Direct the runoff away from sensitive plants and the septic system.

Step 12. Do the same for each household fixture, running the water until it is free of chlorine. Because of the small quantities of water involved, the septic tank will tolerate the drainage. You can speed the process by first emptying the water heater tank. If odor persists or the taste of the water is unpleasant, install a granular activated carbon (GAC) filter at the well-pump outlet.

Step 13. Replace all filters in the system and disinfect associated surfaces with the Clorox solution, followed by a freshwater rinse.

Step 14. To protect the health of those who depend upon the well, have the water tested at least once a year by a state-certified laboratory.

5

Pumps and Related Components

This chapter focuses on conventional, grid-powered pumps and surface hardware. It describes how pumps and distribution systems work, the trade-offs involved, and what to do when something goes wrong, which it will.

Safety

Licensed professionals should install the pump and surface equipment in accordance with the manufacturer's instructions, the National Electrical Code, and applicable local codes.

While any deviation from code increases risk, it is especially critical that equipment be grounded, provided with disconnects, and protected from lightning strikes. Water-well circuitry acts as a magnet for lightning.

115 V AC household current is lethal and never more so when "hot" conductors are within easy reach of grounded well piping. Unless you have the necessary experience, do not attempt to install or service electrical systems.

Well Yield

The primary restraint, the one that we must live with, is the well replenishment rate, or the number of minutes required for the water level in the aquifer to regenerate itself. If the replenishment rate is 0.5 gpm (gallons per minute), then 0.5 gpm is all that we can draw from this well on a continuous basis.

Installing a larger pump costs money and electrical energy, but does nothing to improve the replenishment

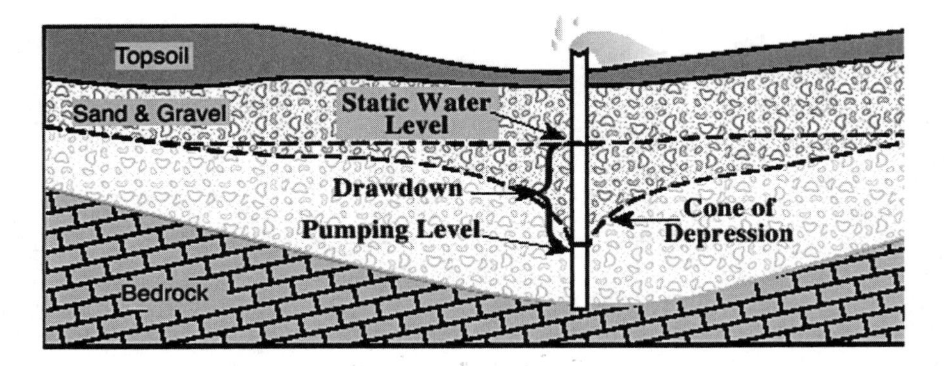

Figure 5-1. *Drawdown sets the limit on the pump size. An overly large pump pulls the water level down below its intake port and quickly destroys itself.*

rate. In fact, the larger pump can run the well dry and, in the process, damage both itself and the reservoir (Fig. 5-1).

Pump Performance

A pump generates flow and pressure. In the United States, flow is expressed as gallons per minute (gpm) or gallons per hour (gph), and pressure as pounds per square inch (psi) or head feet. The latter term refers to the vertical distance the pump can raise water.

> To convert psi to head feet, multiply by 2.31: 10 psi = 23.1 head feet.
>
> To convert head feet to psi, multiply by 0.43: 10 head ft = 4.3 psi.

Total Dynamic Head

The Total Dynamic Head (TDH) is the sum of all downstream resistances to flow. TDH consists of the vertical distance, expressed in feet, from the well water level to the surface, plus any additional vertical distance to the tank, plus frictional loses in the piping, plus tank pressure. The U.S. Department of Energy simplifies things by using the total depth of the well, rather than the water level, as the starting point for the calculation. Since the well is always deeper than its standing water level, the extra lift compensates for pipe friction, which can be ignored.

TDH = Well Depth + Additional Lift to Tank + Tank Pressure

Suppose the well bottoms out at 150 ft, and the tank inlet stands 6 ft above the surface. Tank pressure is 50 psi. Multiplying pressure by the conversion factor 2.31 gives us 115.5 ft of head in the tank.

TDH = 150 + 6 + 115.5 = 271.5 head ft

To be on the safe side we need a pump that develops a TDH of 300 ft.

If you want more flow from an existing well/pump combination, reduce storage-tank pressure or, if practical, straighten and/or enlarge the diameter of the surface piping.

Pressure Regulation

Most residential pressure-controls are factory set to turn on the pump at 30 or 40 psi and shut it down at 50 or 60 psi. Most pumps automatically shut down when discharge pressure reaches 75 psi, and storage tanks have, or should have, pressure-relief valves that open at about the same 75 psi pressure.

Water Demand

Demand varies with geographic location and life style, but it's fair to say that a family of four requires 200 to 400 gallons per day.

Demand peaks during seven-minute periods—in the early morning, as people prepare to leave the house, and in the early evening when they return from work and school (Table 5-1). Peak demand is also a function of the number of bathrooms. The relationship is nonlinear: as more bathrooms (and more occupants) are factored in, the use of washing machines, showers, and toilets increases disproportionately.

Note, however, that the table says nothing about lawn and garden water use, which varies in different parts of the country

Table 5-1
Seven-minute peak demand approximations

Fixture	Flow rate (gpm)	Single Bathroom	1½ to 2 Bathrooms	3 to 4 Bathrooms
		Use during 7 minutes (gal)		
Toilet	4	5	15	22
Lavatory	4	2	6	8
Shower	5	35	53	70
Kitchen sink	5	3	3	4
Clothes washer	5	—	18	18
Dishwasher	5	—	3	4
Seven-minute total		45	98	126
Rate of water delivery		6.5 gpm or 390 gph	14 gpm or 840 gph	18 gpm or 1080 gph

by a factor of 20. Nor does the data provide for hot tubs and swimming pools.

The table assumes a middle-class lifestyle; small, environmentally conscious families can and do get by with much less water.

A 1½ bathroom house can be expected to draw 14 gpm during seven-minute demand peaks. That means that 98 gallons of water (14 gpm times 7 minutes) must be on hand. If the well/pump combination yields 5 gpm, or 35 gallons over seven minutes, some form of storage will be needed to supply the 63-gallon shortfall.

Pump Sizing

Matching the pump to the application depends upon

- well yield, which can vary seasonally,
- Total Dynamic Head,
- peak period demand with more than a passing concern for average daily demand, and
- storage tank capacity.

Integrating these factors requires a sophisticated computer program, such as the one used by Grundig dealers, or experience of the kind accumulated by local drilling contractors, pump suppliers, and state environmental agencies.

Maintenance Issues

Durability data, compiled by the EPA and reproduced in Table 5-2, will give you an idea of the fallibility of pumps, tanks, and other components. Digital motor controls are especially trouble-prone, and have a reported lifespan of only five years.

This data argues for simplicity: adding complexity to the system does not promote reliability or peace of mind. As the Russian arms designer Mikhail Kalashnikov put it, "When a young man, I read somewhere the following: God the Almighty said, 'All that is too complex is unnecessary, and it is the simple that is needed'."

Circuit boards are, in the writer's opinion, among the most egregious examples of counter-productive complexity. Boards tend to fail early, and being proprietary items, unique to each manufacturer, they have high replacement costs.

Table 5-2
EPA durability estimates

Component	Useful Life (in years)
Wells and springs	25
Pumps	10
Hydropneumatic storage tanks	10
Piping	30–35
Valves, manual	35
Valves, solenoid or pressure-controlled	15
Wiring	20
Computerized controls	5
Electronic sensors	7

Piping

Surface piping, that is, piping that runs between the wellhead and the pressure tank, should be level or run slightly uphill, without high spots that collect air. Try to avoid flow-robbing ells and tees. The pressure switch should be on the same level as the storage tank and mounted as close as possible to it.

Standard Jet Pumps

About a third of the 15 million residential wells in the United States use shallow-well jet pumps (Fig. 5-2). These devices consist of a motor, a centrifugal impeller, and a convoluted bit of plumbing known as a jet assembly, or eductor (Fig. 5-3). Water enters at the eye of the impeller and undergoes acceleration as it moves outward towards the impeller rim; the bulk of it exits through the discharge port. Some portion of output, known as drive water, recirculates back through the jet assembly.

The water in the well is under atmospheric pressure. As the drive water passes through the venturi, it speeds up and, at the same time, loses pressure. Well water, responding to the pressure differential, flows into the pump. Because atmospheric pressure is only 14.7 psi, a pump that develops a perfect vacuum at its intake port would lift only 34 ft. The partial vacuum generated by real-world jet pumps limits lift to about 20 ft.

Jet pumps provide an inexpensive and reasonably reliable means of extracting water from shallow depths. Surface mounting—either on the wellhead or offset some distance from it—is a great advantage, especially for homeowners who do their own maintenance. Some manufacturers market integrated systems with the pump, pressure tank, valves, and controls already plumbed and wired. Such consumer-oriented products simplify installation. Better systems can be built by shopping around for quality components and assembling them yourself.

Pump output and pressure depend upon motor horsepower and well depth. For example, the Sta-Rite HCL ½-hp unit delivers 11.7 gpm at 30 psi from a 15-ft well. If we increase the pressure to 50 psi, output drops to 6.1 gpm. Increasing well depth to 20 ft reduces output by 25 percent.

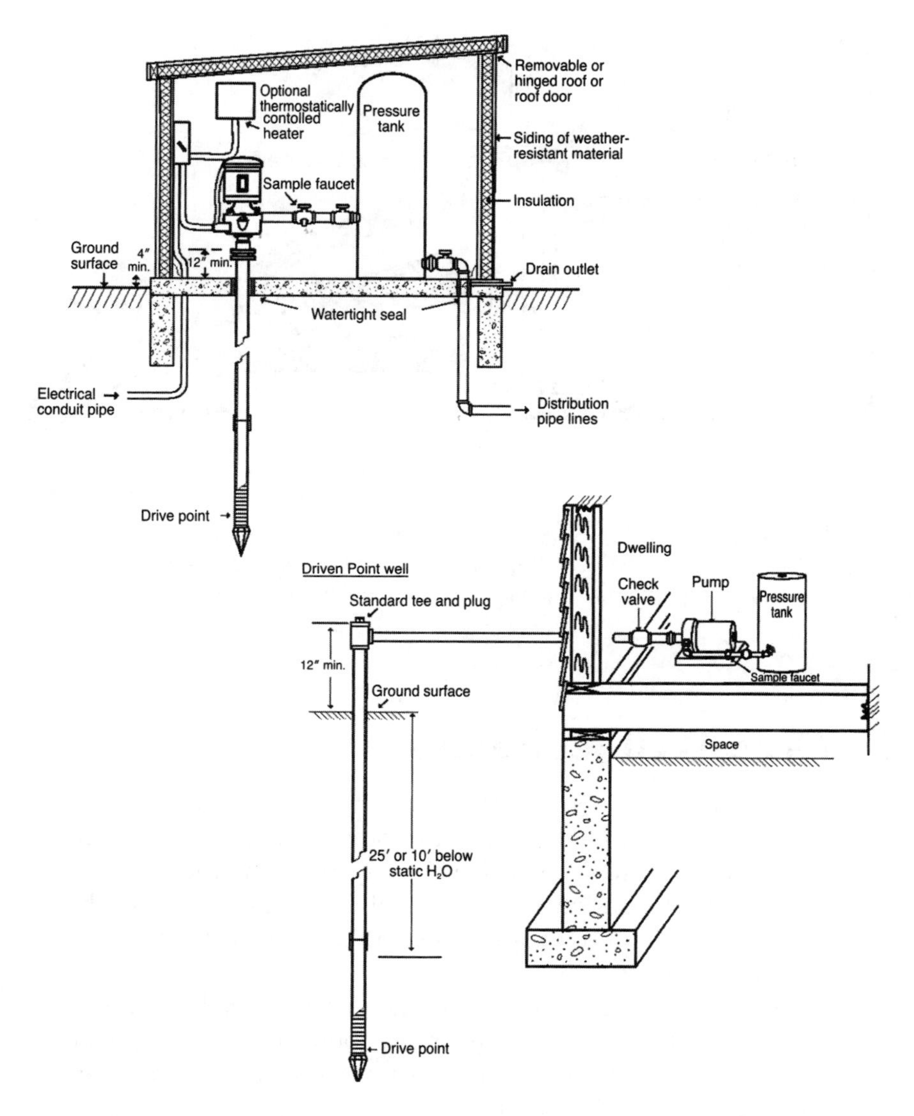

Figure 5-2. *A jet pump mounts either directly on the wellhead or at some distance from it. Both installations require a downhole foot valve (not shown) that closes to prevent water from draining out of the drop pipe when the pump stops. An offset pump has two provisions for priming—a priming plug on the pump casing and a pipe plug on the drop pipe tee. A check valve just downstream of the pump intake assures that the impeller retains prime.* Wisconsin Department of Natural Resources

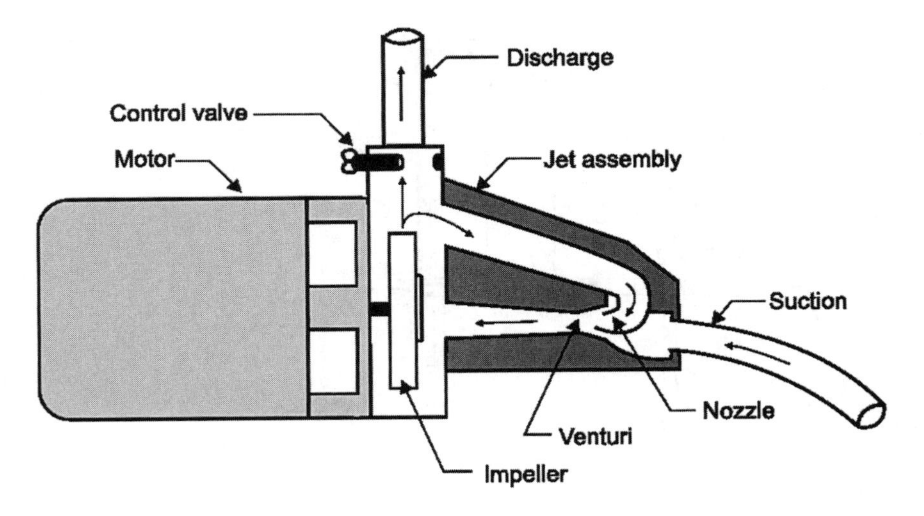

Figure 5-3. *A jet pump schematic. The venturi increases the velocity of the drive water stream while reducing its pressure.* Government of Alberta, Agriculture and Rural Development

Deep-Well Jet Pumps

Separating the jet assembly from the motor and impeller enables the pump to be used at depths of 70 ft or more. This is possible because the jet assembly is submerged, which reduces its dependence on atmospheric pressure. Now the pump "pushes" rather than "pulls." Most deep-well jet pumps peak out at around 70 psi.

As shown in Figure 5-4, the submerged jet assembly receives drive water through a second pipe concentric with the well casing. To accommodate these components the casing must have a diameter of at least 4 inches.

A tail pipe can be added to extend the pump intake below the level of the jet assembly. When this is done, suction again comes to the fore, and the longer the tail pipe, the less efficient the pump.

Refer to Table 5-3 for troubleshooting shallow- and deep-well pumps.

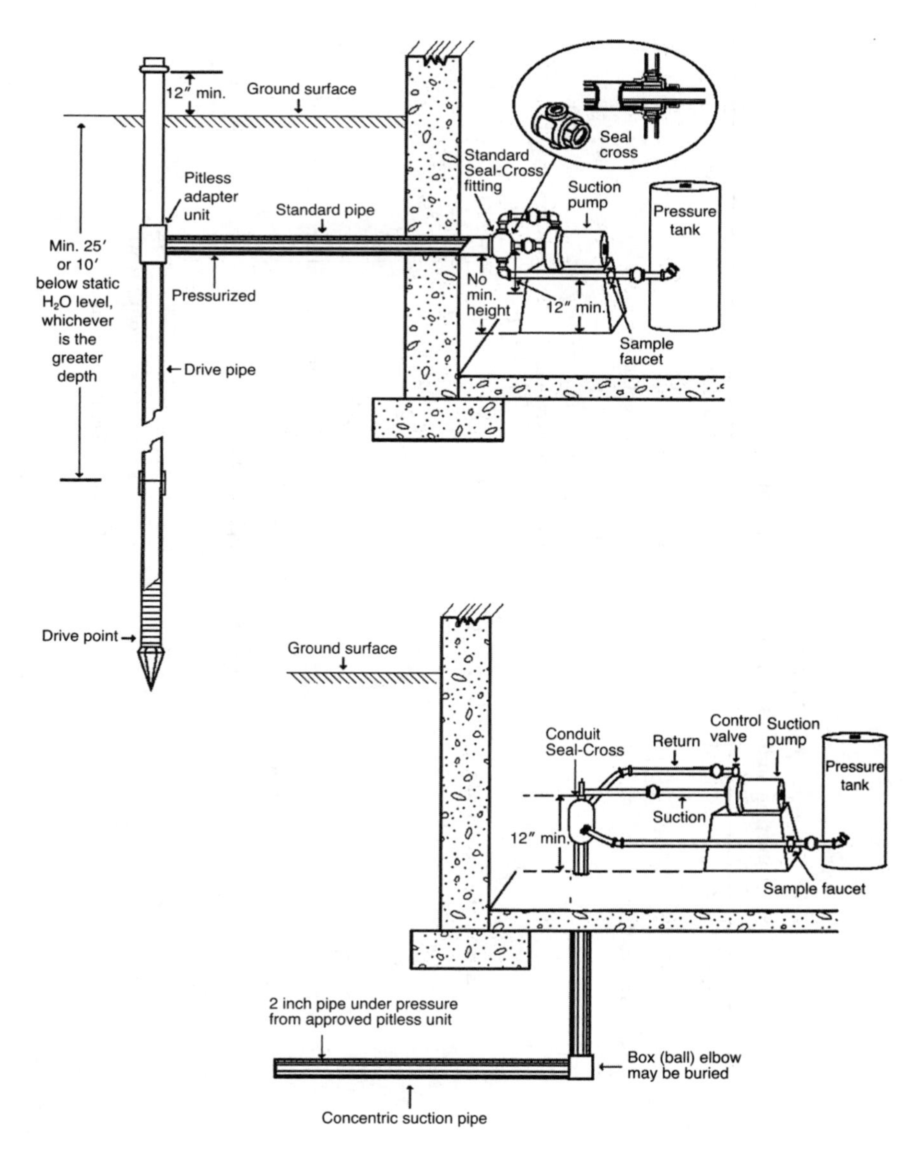

Figure 5-4. *Deep-well jet pump installations. The buried horizontal piping provides freeze protection. A foot valve (not shown) mounts at the water intake and automatically closes when the pump stops to maintain prime.* Wisconsin Department of Natural Resources

Table 5-3
Troubleshooting shallow- and deep-well jet pumps

Symptom	Probable Cause	Corrective Actions
Pump motor does not run	Fuse blown or circuit breaker tripped	Restore power.
	Motor thermal overload switch opened	Wait, sometimes 30 minutes or more for the motor to cool and the thermal switch to reset itself. Try to determine the cause of overheating, which usually can be traced to short cycling. Thermal switches on some pump motors must be manually reset.
	Starting capacitor failed	Test and replace.
	Motor switch failed	Test and replace.
	Pressure switch failed	Test and replace.
Pump short cycles or runs continuously	Storage-tank air charge low	Recharge the tank.
	Pressure switch incorrectly adjusted	Adjust the switch to the pump-maker's specifications.
	Pressure switch contacts sticking	File contacts for a short-term fix or replace the switch.
	Pressure tank is too small for water demand	Upgrade the tank.
	Faucets open	Close all outlets.
	Piping leaks	Identify and repair leaks, with attention to the "phantom" toilet-tank leaks into the bowl.
	Foot valve leaks	Clean or replace the foot valve.

Complete loss of prime—the motor runs, but no water flows. Shut off the motor immediately. Unscrew the priming plug. If no water is present in the priming hole, the pump has lost prime.	Water level drops below pump lift capacity.	20 ft is the safe maximum depth for shallow well pumps.
	Foot valve or check valve is stuck	Replace the faulty valve.
	Foot valve or well screen is buried in mud or sand	Clean the well or raise the suction pipe.
	Well screen clogged	Retrieve and clean the strainer.
or		
Partial loss of prime—sputtering faucets; however, some sputter and "hiccupping" may be present upon recharging the pressure tank.	Suction pipe leaking air	Check with soapy water and compressed air.
	Pump eductor or impeller clogged	Clean parts.
	Well overpumps, that is the water level periodically falls below the foot valve	Lower the foot valve if possible, otherwise limit the amount of water used.
	Gaseous well	
Pump fails to deliver rated flow	Filter(s) clogged	Replace .
	Pump partially clogged	Disassemble and clean.
	Steel pipe corroded	Replace with PVC or poly .
	Low water level	20 ft is the safe maximum for a shallow-well pump.

Cavitation

Cavitation occurs when the pump is mounted too high or if the pump lowers the water level more than anticipated. The impeller creates a vacuum that enables air and gas bubbles to form. In other words, the water goes into a kind of cold boil. The bubbles attach themselves to the impeller and pump casing, where they implode violently. Each implosion leaves a pit on the metal surfaces. Water delivery becomes sporadic. Severe cavitation sounds as if marbles were rattling around inside the pump. Mild cavitation may not be heard, but over time has the same destructive effect.

The impeller and venturi throat—the narrowest part of the inlet track—takes the brunt of the cavitation damage. Sand-induced abrasion will be most pronounced at the rotor tips. Rust bubbles and wide, shallow pits on the cast-iron parts, or a film of white oxide on aluminum, point to corrosion damage.

To make a definitive diagnosis, measure the pressure on the suction side of the pump with a combination vacuum/pressure gauge available from auto parts stores. Mount the gauge as close as possible to the pump inlet. Consult the pump manufacturer's literature to determine the Net Positive Suction Head (NPSH). If suction pressure drops below NPSH, the pump will cavitate.

Shaft Seal Replacement

Water leaks at the joint between the pump casing and motor housing mean that the pump seal—the seal between the impeller and motor—has failed. Sand, corrosion, or a chronic loss of prime make replacing the seal a routine chore. Should leaking water short out the windings, you will need to purchase a motor, even if the pump is new. Warranties do not cover water damage.

Mechanical shaft seals, also known as face seals, consist of a ceramic stationary element and a carbon or elastomer rotating element, pressed into rubbing contact by a coil spring.

To replace the seal:

Step 1. Remove the pump housing and the terminal cover at the rear of the motor.

Step 2. Impellers thread onto the pump shaft. Secure the motor shaft with a screwdriver inserted into the slot provided (Fig. 5-5). Using a strap wrench, unthread the plastic impeller. Most motor shafts have the standard "right to tighty, left to loosey" threads. Motors that rotate counterclockwise when viewed from the impeller end nearly always have a left-hand shaft thread.

Step 3. The face seal is accessible with the impeller removed (Figs. 5-6 and 5-7).

Step 4. If not damaged, the pump-casing o-ring or gasket can be reused.

Fig. 5-5. *The motor shaft is slotted for screwdriver purchase. A piece of leather prevents damage to the plastic impeller as it is unscrewed from the shaft.* Tony Shelby

Figure 5-6. *Once the impeller is unscrewed, the shaft seal is accessible.* Tony Shelby

Priming

To prime, follow this procedure:

Step 1. Unscrew the priming plug located on the pump casing or on a tee at the suction line. Some older pumps use a pressure gauge as the priming plug.

Step 2. Fill the surface suction line and pump casing with water.

Step 3. Close the gate valve at the pump outlet.

Step 4. With the plug slightly loose, start the pump.

Step 5. Shut off the pump as soon as a steady stream of water flows out around the priming plug.

Step 6. Snug down the priming plug.

Step 7. Open the gate valve and re-start the pump.

Figure 5-7. *The face seal consists of a carbon-lined rotating face and a ceramic stationary face. Most pump casings have an o-ring seal, as shown in the right of the photo. These seals can generally be reused.* Tony Shelby

It may be necessary to recharge the storage tank with compressed air. (See the "Recharging" section in this chapter.)

Centrifugal Submersible Pumps

Centrifugal submersible pumps (sub pumps or subs) rated at between 5 and 10 gpm are the overwhelming favorite for homeowners with wells too deep for jet pumps (Fig. 5-8). Almost all of these motors require a minimum casing diameter of 4 in. Smaller, 3-in. subs exist, but are expensive.

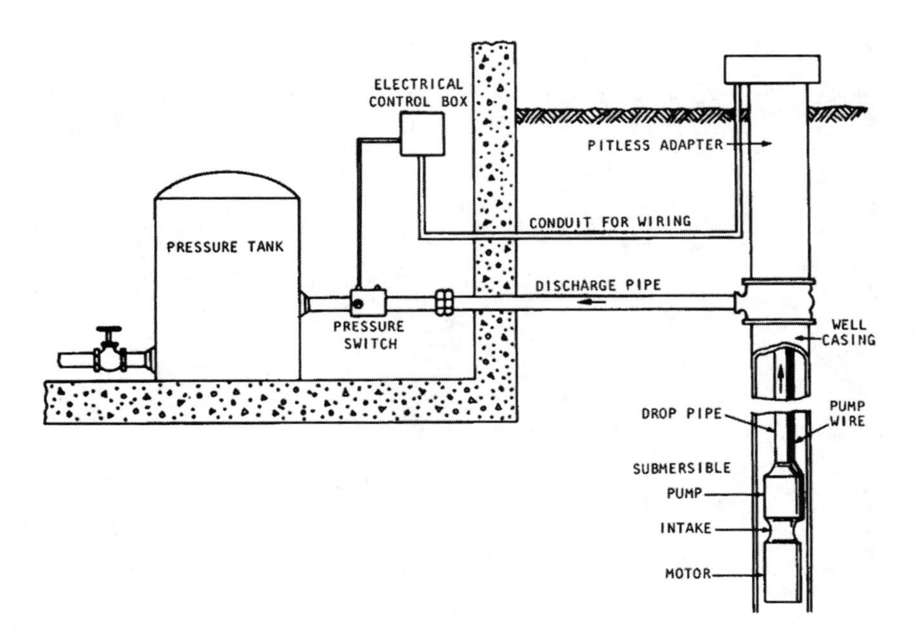

Figure 5-8. *A typical sub pump installation.* Illinois Department of Registration and Education

Figure 5-9 illustrates the parts arrangement for an oil-cooled sub pump. Other, generally less popular, designs cool the motor with water. While oil does not transfer heat with the facility of water, it does provide lubricity and some protection against short circuits. However, be certain that the lubricant is "food grade."

Few subs run for 30 years. But if you are unfortunate enough to have an oil-filled sub built before the early or mid-80s (no one is sure of the cutoff date), have the water tested for PCBs. Some, probably all, of these subs used what was known in Vietnam as Agent Orange for cooling and as a start and run capacitor dielectric. And, like any sub, they leaked. More information about PCB-contaminated subs can be found at http://dnr.wi.gov/org/water/dwg/gw/pcb.pdf.

Regardless of the cooling provision, a 3450-rpm induction motor mounted at the base of the sub drives the pump

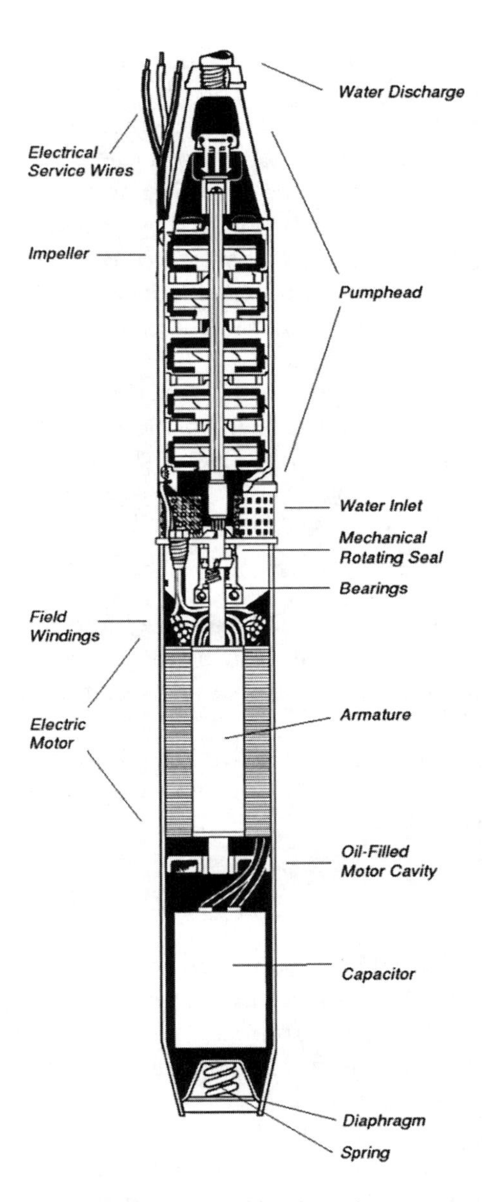

Figure 5-9. *An oil-filled centrifugal sub with an integral starting capacitor. The spring-loaded diaphragm maintains oil pressure at a value greater than the pressure of the surrounding water. Some, hopefully food-grade, oil finds its way into the water stream. Water-cooled motors have better heat transfer than their oil-cooled cousins, but are harder on the bearings.* Sandia Laboratories

through a splined shaft. Splines isolate pump thrust from the motor bearings. The pump assembly, or "bowl," consists of stacked centrifugal pumps, each of which makes up a stage. Subs sized for 4-in. casing generate about 9 psi of head pressure per stage. When more head is needed, manufacturers simply add more stages. Modular construction accounts, in great part, for the relatively low cost of these units.

For durability some manufacturers specify 300-series stainless-steel impellers. Others use glass-reinforced plastic. But neither material is proof against sand.

Water-lubricated thrust and axial bearings support the pump shaft. Thrust is the big problem, since down force increases with each foot of pressure head. Designers make the lower thrust bearing as large as space permits and further reduce wear with lift-generating impeller geometry. Even so, the lower thrust bearing usually is the first part of the pump to wear out.

During starting, when well-water levels are high, the reverse situation occurs. There is less head pressure and the lift-inducing impeller stack presses against the smaller upper thrust bearing.

Because the upper thrust bearing is designed for only transitory loads, chronic up-thrust—that is, using a pump with too many stages for the TDH—is a leading cause of pump failure.

Wiring

Most subs require 115 V single-phase alternating current or 230 V single or three-phase AC. Some are also compatible with 208 V three-phase current, an oddball voltage sometimes encountered in nonresidential applications. But whatever the voltage, having three-phase current reduces power bills and makes for a more reliable motor.

Wiring nomenclature is a bit misleading. A two-wire pump requires three wires—red, black, and a green ground. A three-wire unit has four wires—red, yellow, and black for power and a green ground wire. The two-wire configuration is limited to smaller pumps with integral starting circuitry. This configuration simplifies the connections, and, of course, requires less wire. But should something go wrong with the starting circuitry, the pump must be pulled.

Three-wire systems operate from a surface-mounted control box. The box contains starting and run capacitors and the starter relay, and may include variable-speed/constant-pressure controls, soft stop/start circuitry, and protection against dry running. The last feature is particularly important. Three-wire circuitry also provides high starting torque, an advantage in silty wells.

Installation

While installation should be put in the hands of a professional, there are aspects of the job that homeowners should be aware of. One of these aspects is the type of piping used.

The choice of drop-pipe material depends upon local practice and the driller's preferences. While galvanized steel has lost ground to plastic pipe, many drillers would use nothing else for wells deeper than 200 or 300 ft. Others use Schedule 40 PVC exclusively, in spite of evidence that this light-gauge tubing is unsuitable for even moderately deep wells. Schedule 80 or 120 PVC are better, although more expensive, options.

Heavy-wall PVC comes in 20-ft lengths with female pipe threads at both ends. Sections are joined with male couplings. Stainless steel couplings are the safest choice for deep wells, although machined or extruded PVC couplings have also given good service. Lead-free brass—if you have confidence that the material is actually lead free—may also be used.

High Density Polyethylene (HDPE or "poly") came on the market in the 1960s and has since become the standard for municipal water lines. Poly was also the material of choice for the 79,000-ft fire-suppression-piping network at the nuclear weapons plant near Amarillo, Texas. A more critical application could hardly be imagined. Grade PE 3408 has become increasingly popular for drop pipe.

Poly is flexible—capable of being bent in a radius 25 times its diameter—and comes in coils the distributor cuts to length. It installs in a continuous operation without the need to stop every 20 ft or so to make up another section. Life estimates range from between 50 and 100 years: one manufacturer puts no time limit on its warranty.

The inside diameter (ID) is constant for each size of pipe. The pressure the pipe withstands depends upon the wall thickness, expressed as SDR. The Standard Dimension Ratio is the ratio of ID to wall thickness. The lower the number, the thicker the wall, and the more pressure the pipe can withstand.

To calculate pressure requirements we need to know two things: the maximum water depth, a figure that takes into account the drawdown when the pump runs, and the maximum pressure setting of the storage-tank switch. Since the pump must develop 0.433 psi for each foot of head, the total pressure generated is equal to:

$$\text{water depth} \times 0.433 + \text{maximum tank pressure in pounds per square inch.}$$

The calculation does not take pipe friction into account.

Thus, if the water depth is 360 ft and the pressure switch turns the pump off at 40 psi, the drop pipe must be able to withstand:

$$360 \times 0.433 + 40 = 195.9 \text{ psi.}$$

Poly pipe rated at 200 psi or higher will serve our purpose.

The driller will handle details such as the number and spacing of check valves, the method of securing the power cable to the pipe, the type of connections used to make wiring splices, and whether the safety line is polypropylene or stainless steel. Nylon should not be used, since it absorbs water and weakens.

Durability

A centrifugal sub should return 15 years or more of service. Assuming that the pump is sized for well yield and TDH, the major causes of premature failure are:

- **Sand**—A sand separator, sold separately, will block most sand entry.
- **Frequent starts**—The surge of current needed to put the motor into motion is death on insulation and electronic control circuitry.

- **Voltage surges**—Lightning strikes, inadequate wiring, and poor voltage control by rural cooperatives all contribute to this problem.

Water Storage

The well casing acts as a storage tank and, depending upon diameter and depth, can hold a considerable volume of water (Table 5-4).

It is good practice to specify casing two sizes larger than the minimum required by the pump diameter. Going from 4-in to 6-in casing more than doubles the storage capacity. Depending upon casing size, well depth, recharge rate, and water needs, it may be possible to rely entirely upon casing storage. However make dead certain that your water requirements are moderate and will not short cycle the pump with frequent starts. And install a low-water-level cutoff switch.

A storage tank connected by way of a booster pump to a pressure tank makes it possible to live comfortably with a low-yield well. The well pump replenishes the storage tank over 12 hours with enough water to meet early morning and evening requirements.

A cistern is an underground or surface reservoir with a removable cover to facilitate cleaning and disinfection (see Chapter 3). Most have capacities of 1500 gallons or more.

Table 5-4
Casing volume

Casing Diameter (in.)	Gallons of Water per foot
3	0.37
4	0.65
5	1.0
6	1.5
8	2.6

Older cisterns were almost always made of concrete; the modern preference is for steel or plastic. If you opt for plastic, investigate the formulation because health hazards may exist. A booster pump, sometimes submerged in the tank, moves water on demand to a pressure tank. Several thousand gallons of stored water provide a cushion when the well stops working (as it will sooner or later) and comes in very handy in the event of fire.

Gravity Tanks

Gravity tanks have the virtue of simplicity, but impose siting problems if normal pressure is the goal. To develop a delivery pressure of 40 psi, the tank must be elevated 90 ft above the outlet. However, urban residences in the less affluent parts of world get by nicely with roof-mounted gravity tanks. One quickly becomes accustomed to the limited pressure. And, of course, it is always possible to install a booster pump.

Captive-Air Tanks

Captive-air, or hydropneumatic, tanks store water under air pressure. Sales literature describes these tanks in terms of the number of gallons they would hold if empty of air. But air, even when compressed, occupies space. The critical parameter is drawdown volume, or the amount of water the tank can deliver before the pump starts. One popular 50-gal tank has a drawdown volume of 18.3 gal when operated between 20 and 40 psi. At 40 to 60 psi, the drawdown volume shrinks to 13.4 gal.

Follow this procedure to measure drawdown volume for captive-air tanks of all types:

- Open a water spout to drain the tank.
- When the pump starts, close the spout and place a 5-gal bucket under it. If you cannot hear the well pump run, have someone report when the pressure-switch contacts click closed.

- The pump will stop when the cut-off pressure is reached. At that point, the tank holds as much water as the system permits. Open the spout.
- When the pump restarts, the amount of water caught in the bucket is the drawdown volume.

Bladderless

Some hydropneumatic tanks have no diaphragm to separate air and water. Eventually, pressure leaks off as the air dissolves in the water. In their most basic form, these tanks have no means of automatic air replenishment; every few months or so, the owner has to pump up the tank.

Figure 5-10 illustrates how a small, diaphragm-type compressor maintains tank pressure. A water-level sensing switch

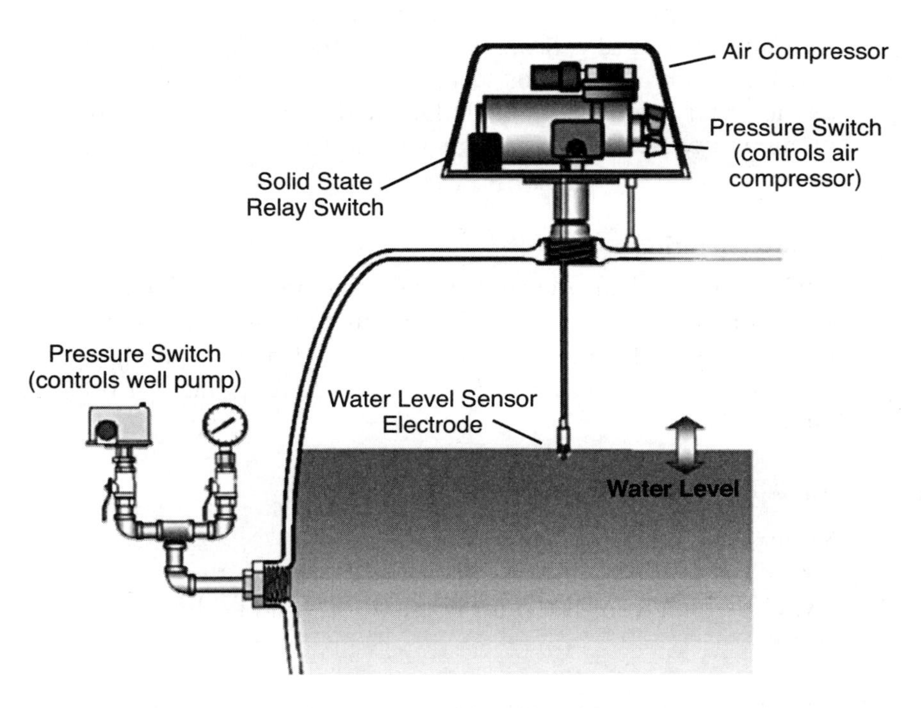

Figure 5-10. *State-of-the-art bladderless tanks employ a small compressor to maintain the air charge.* Washington State Department of Health

and an air-pressure switch work in conjunction to turn the compressor on and off. The pressure switch on the left of the drawing controls the well pump.

The next drawing, Figure 5-11, illustrates a clever arrangement for recharging that does away with the air pump. When the sub pump stops, the bleeder valve mounted in the drop pipe opens to drain the water standing above it. A check valve closes to isolate the tank. At the same time, the sniffer valve, mounted at some accessible location on the horizontal piping, opens to admit air. When the pump comes on, bleeder and sniffer valves close and the check valve opens. Entering water compresses the air in the tank.

While this system has the advantage of simplicity, the sniffer and bleeder valves are perennial sources of trouble. Replace them at the first sign of short cycling.

A third approach to air charging employs a Schrader valve to admit air into the drop pipe when the pump stops (Fig. 5-12). A float valve opens to bleed excess air from the tank as the water level is replenished. Note that the Schrader valve element does not interchange with auto tire valves of the same name.

Troubleshooting

Short or rapid pump cycling usually means that the tank is water logged as the result of loss of air pressure. An application of soap and water will reveal external leaks. If the piping is sound, possible causes of low or no air depend of upon tank design.

Tanks with air compressors.

Verify that the compressor receives power. Using a voltmeter or test lamp, trace the circuit back through the pressure switch (which turns the pump on and off) and to the breaker or fuse.

Warning: *At the risk of repeating myself, I want to emphasize the dangers of electrical work in a grounded environment. Should one be caught between 115 V power and a well pipe,*

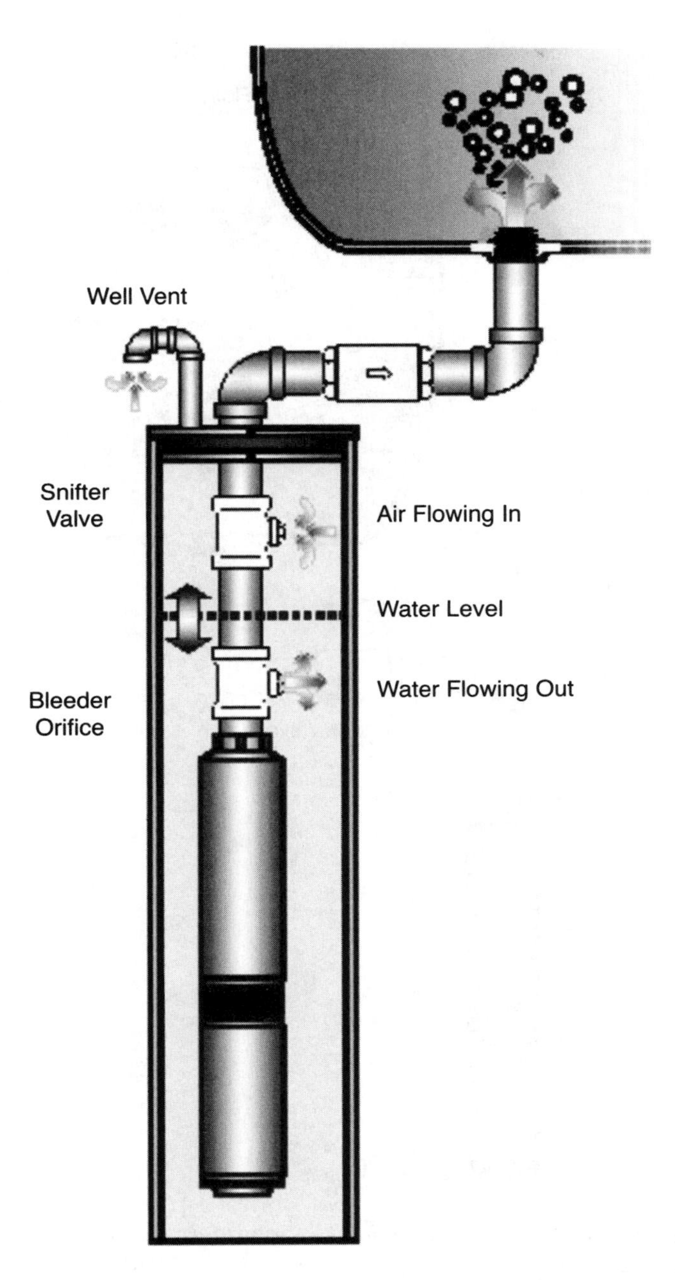

Figure 5-11. *A bladderless tank with the infamous bleeder and sniffer valves.* Washington State Department of Health

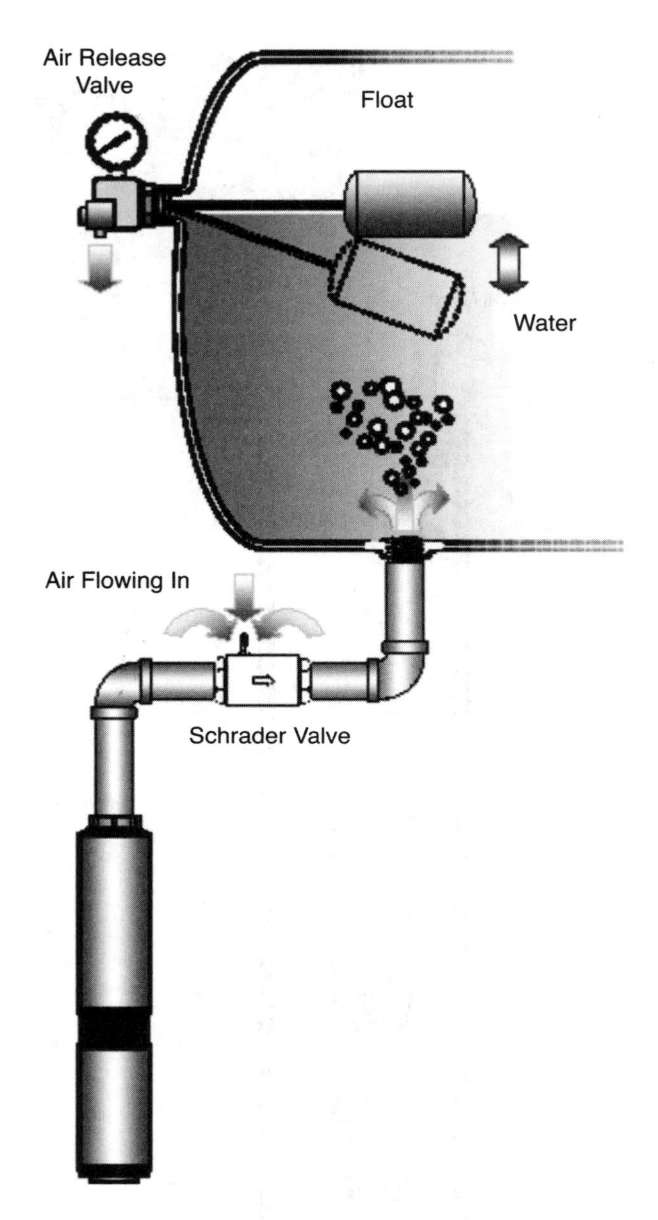

Figure 5-12. *In this arrangement, a Schrader valve mounted on the surface piping opens to admit air. The float valve opens to bleed overpressure.* Washington State Department of Health

the result could be lethal; 220 V removes all doubt about the outcome. If you are inexperienced or uncomfortable around electricity, hire a professional to make these tests. And be aware that wells, because they are such excellent grounds, attract lightning. Do not work in stormy weather.

If the pump receives power, open the breaker and disassemble the pump far enough to be able to turn the crankshaft by hand. Resistance should build near the top of the stroke as the diaphragm is displaced. Few of these pumps are repairable if the bearings have gone. Replacement diaphragms can sometimes be found.

Excessive air pressure that sends water gushing out of the spouts or trips the pressure relief-valve means that the pump-control pressure switch contacts have welded themselves closed. Should both the pressure-relief valve and the pressure-switch fail, the tank may rupture.

Tanks that utilize ambient air pressure.

The valves—bleeder, sniffer, or Schrader—live in a hostile environment. Bleeder valves are fairly robust, since the downhole pump generates closing pressure; sniffer and Schrader valves are delicate contrivances that open and close by atmospheric pressure. In the absence of tank air pressure, blame the atmospheric valve that should be mounted on the surface piping for easy access. If replacement does not solve the problem and no obvious leaks are present, lift the drop pipe high enough to replace the bleeder valve.

Should the tank become over-pressurized, replace the air-release valve.

Diaphragm

Diaphragm tanks have become almost universal for residential systems. A flexible diaphragm—also known as a bladder or membrane and, when spherical, as an air cell—separates the air chamber from the water (Fig. 5-13). A pressure switch controls the pump motor. Water enters the bottom of the tank past a check valve. The air chamber carries a slight precharge.

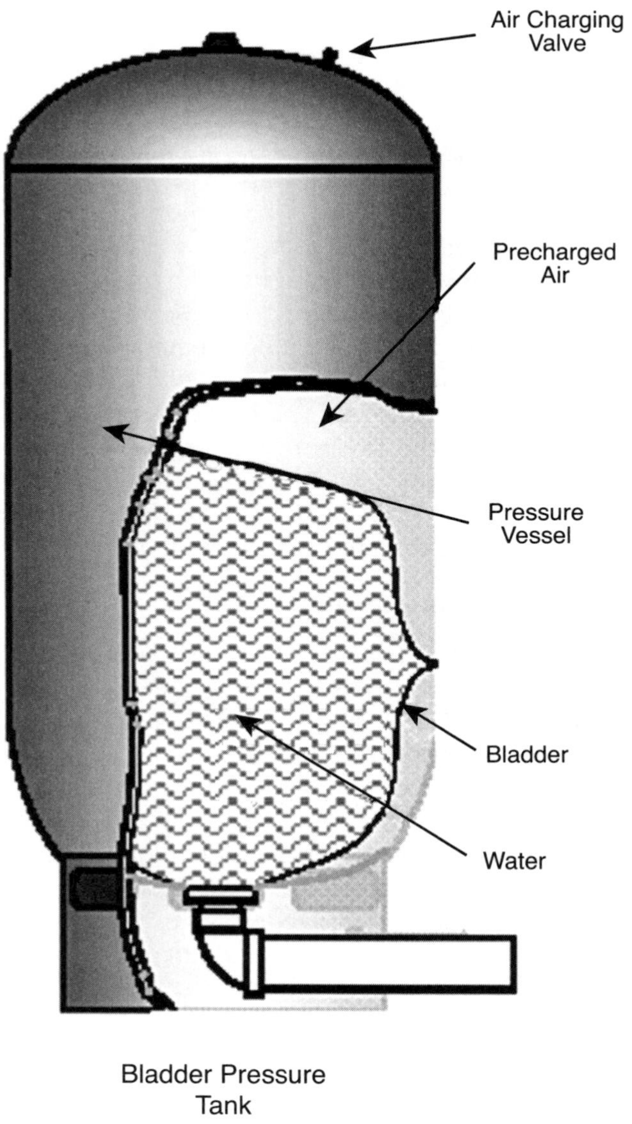

Figure 5-13. *A diaphragm-type pressure tank.* Washington State Department of Health

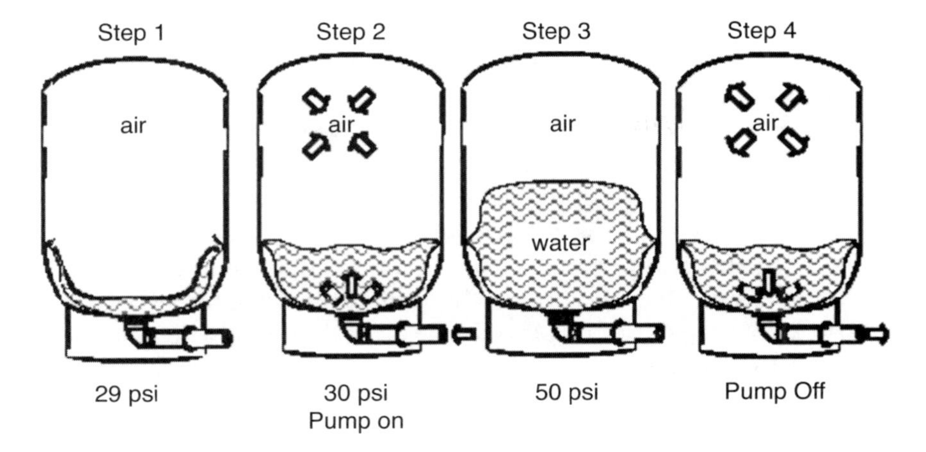

Figure 5-14. *Bladder tank operation.* Washington State Department. of Health

Figure 5-14 illustrates the inverse relationship between air and water volumes. In Step 1 of the drawing, the tank is empty, and the diaphragm, responding to the 29-psi precharge pressure, elongates to occupy most of the available volume. Note that the precharge pressure on the diaphragm must be a few psi lower than pump motor cut-in pressure. If precharge pressure were higher, the motor would not start.

The pump starts and begins to fill the tank (Step 2). As the water level rises, the diaphragm expands to reduce the volume of the air chamber. Air and water pressure both increase to, in this case, 50 psi (Step 3). The pressure switch then opens to shut off the pump. As water is used, pressure falls until the 29-psi cut-in value is reached, and the pump again comes on.

Recharging

In a perfect world, precharge air pressure would remain constant over the life of the tank. However, some leakage is inevitable, and every few years the residual air charge dissipates in otherwise healthy systems. When this happens the tank becomes waterlogged. Without the "spring" compressed air provides, the smallest draw on water drops the pressure enough to energize the pump. Since water behaves as a solid, pressure builds almost instantly, and the pump stops. The pump motor cycles in this manner until it burns out.

If you do not know the precharge pressure, ask the vendor or a local plumber. Most tank manufacturers include this data on their web sites. In the absence of information to the contrary, charge the tank to 2 or 3 psi below pump cut-in pressure. If the pump starts at 30 psi, the residual pressure—the air pressure that remains in the tank after draining—should be 27 or 28 psi. But this is only a rule of thumb; always check with the manufacturer or vendor. Some large, and presumably more vulnerable, diaphragm tanks have a precharge-pressure limit of 22 psi.

To recharge,

- Disconnect the power to the pump.
- Open a tap and allow the tank to drain dry. Things go more easily if water remains in the plumbing. A completely dry system will almost surely leak air.
- Close the tap.
- Check the pressure at the Schrader valve on the top of the tank with a tire gauge. If loss of residual pressure has occurred, the gauge will read something considerably less than cut-in pressure.
- Recharge with a hand pump, stopping frequently to check the pressure. **Warning:** Over-pressurization can explode the tank. Unless you are an experienced hand, use a manual pump—not an air compressor. In any event keep a running check on the gauge reading.
- Power up the pump.
- Test by opening several water taps. Ideally the pump should not cycle more than six times an hour.

If the tank loses pressure rapidly, that is, within a few weeks of recharging, suspect a ruptured diaphragm. Most tanks cannot be disassembled to make the repair.

Pump Motor Controls

Pumps, tanks, and controls can be purchased as integrated packages or built up in piecemeal fashion. Integrated systems simplify the installation. But, as mentioned earlier, readers with DIY skills can come out ahead by fabricating their own motor control and valve arrangements. It's not rocket science, and one can use industrial rather than consumer-quality components (Fig. 5-15).

Pump Control and Plumbing System Component List

1. **Breaker box**—provides over-amperage protection for the motor and other electrical components.
2. **Lightning arrestor**—blocks voltage surges caused by lightning.
3. **Control box**—available in 115 V and 230 V versions, the control box protects the pump motor from over- and under-voltage, and rapid cycling. The box also includes

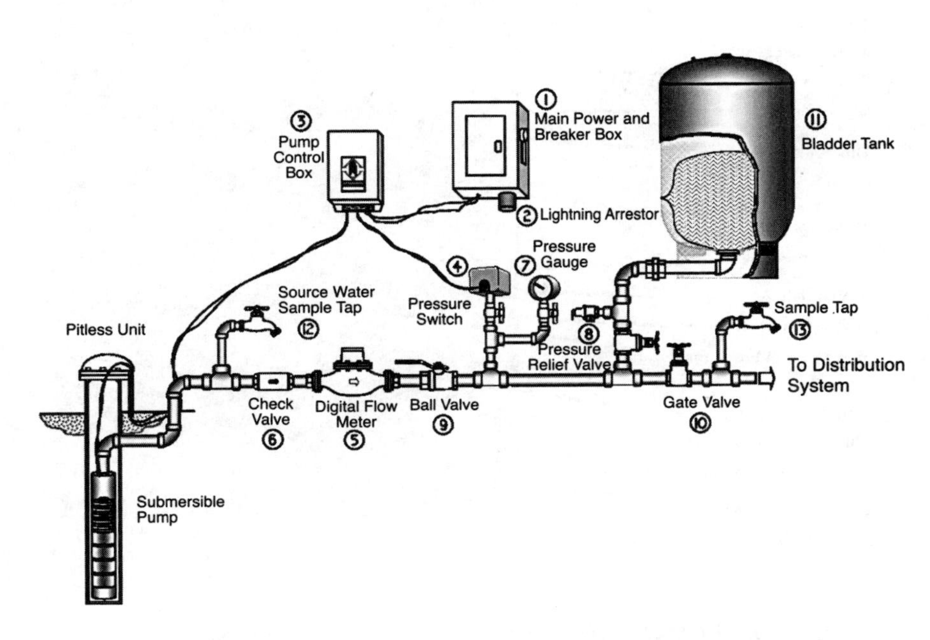

Figure 5-15. *Pump control and plumbing systems should, at the minimum, include the components listed. The electrical layout shown is for a three-wire sub; two-wire subs require the same plumbing arrangement.* Washington State Department. of Health

start and run capacitors, vulnerable components that require fairly frequent replacement.

4. **Pressure switch**—responds to storage tank pressure to turn the pump on and off.
5. **Flow meter**—some states, for example, Minnesota, require that wells be fitted with a flow meter as a means of conservation. The assumption is that we use less of what is measured.

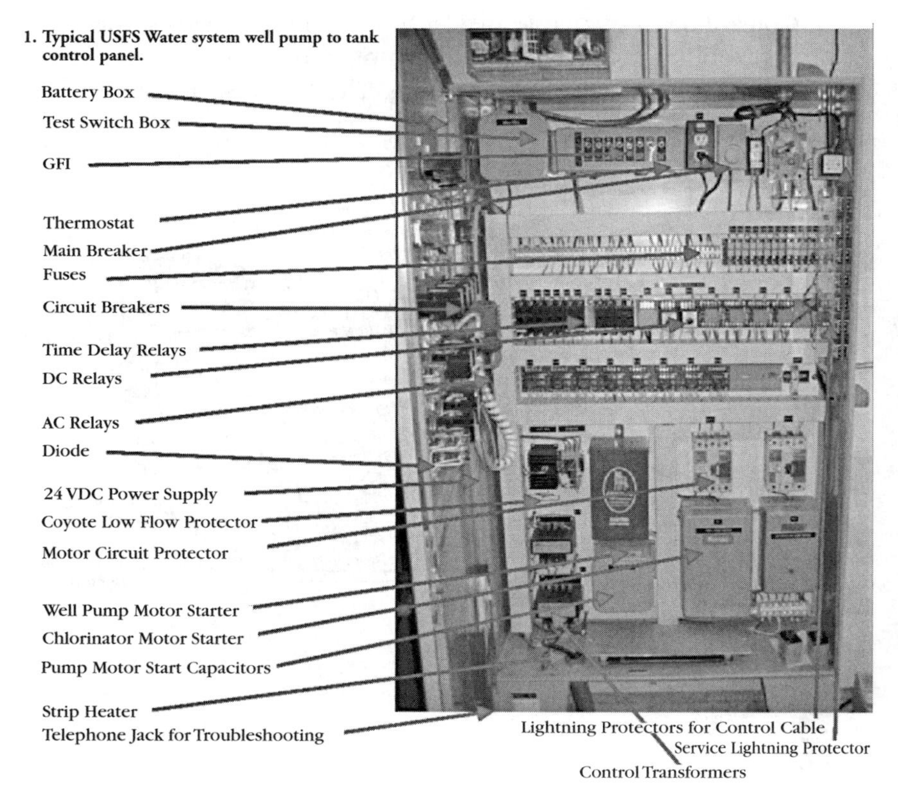

Fig. 5-16. *A USDA Forestry Service control box integrates multiple functions into a single, accessible package. The use of discrete components simplifies diagnosis and repair.*

6. **Check valve**—a one-way valve that prevents the tank from draining back into the well.
7. **Pressure gauge**— reports water pressure.
8. **Pressure-relief (PR) valve**—a safety-critical component that opens to relieve excessive water or air pressure. The valve should be protected from the weather with its discharge directed away from people or electrical components. The PR valve should be one approved by the tank manufacturer.
9. **Ball valve**—normally this valve remains open until the system requires servicing. A ball valve has less of a tendency to stick than the more common gate valve.
10. **Gate valve**—another and sometimes welcome aid to servicing.
11. **Bladder-type storage tank**.
12. **Sample valve**—used to test well-water quality.
13. **Optional sample valve**.

Figure 5-16 illustrates a professionally designed control box, including its components grouped by function, its accessible wiring connections, and a test switch box that simplifies diagnosis. There are no circuit boards.

6

Alternative Power

This chapter describes how solar, wind, and human power are used to lift water. Solar and wind also provide electricity for millions of people throughout the world, many of whom have no other source. Nor is the appeal confined to less developed countries. Installed and maintained correctly, these systems are more reliable than the American grid. According to the Eaton Blackout Tracker, 2,169 power outages, each of which affected 50,000 or more people, occurred in 2008. In 2011, there were more than 3000 outages. And the forecast is for more to come as utility investment lags and the climate becomes increasingly unstable. Alternative

Further Notes on Safety

Solar systems appear innocuous, with none of the noise and intimidation associated with conventional power plants. But electricity, no matter how generated, is to be respected. If you are unsure of your electrical skills, do not try to install a solar system yourself. Hire a licensed professional.

If you feel confident enough to take on the job, obtain a copy of the National Electric Code (NEC) and read Article 690, sections A to H, with the care that a lawyer gives to a contract. This material describes the components that make up the system and how they are to be connected. Section 480 goes into detail about storage batteries. It is also important to verify that panels, controllers, and other hardware carry the Underwriters Laboratories seal of approval.

Take every precaution not to become a conductor between a "hot" wire and a ground wire. Using a voltmeter, measure the electrical potential of switch boxes, BX cable, solar-panel frames, and other metal parts to ground.

Wear dry leather gloves and rubber-soled shoes; whenever possible, work with one hand to keep your heart and lungs out of the circuit. The most commonly encountered solar modules generate about 22V and a 3A open-circuit. A 90mA (0.09A) DC current passing through the chest is enough to produce asphyxia. The victim may not be able to let go of the conductors. Exposure to a 500mA (0.5A) current means probable death; to a 1A current removes the uncertainty. AC amperage is even more lethal.

Anytime a PV panel contains more than two modules, consider it lethal.

In addition to the danger imposed by the electrolyte, lead-acid storage batteries release hydrogen gas during charging. Battery explosions are not uncommon. Shorting a battery, as with a pair of pliers across the terminals, releases as much as 8000 A, or more than enough current to weld railroad track.

power, if only in the form of an emergency backup, is good to have.

Alternative energy is also a way around the high hookup charges imposed by many rural co-ops. If you build in a remote area, you can be asked to pay $5 a foot for an extension to existing power lines. That works out to more than $25,000 a mile.

Except for human power, which is taken for granted, states, local utilities, and the federal government have created incentives for these off-grid technologies. To find out more, go to the Database of State Incentives for Renewables and Efficiency at http://www.dsireusa.org.

Solar Power

If 2.6 percent (155,000 square miles) of the Sahara Desert were covered over with solar panels, enough electricity would be produced to satisfy the needs of the entire planet. One hour of the solar radiation that strikes the Earth could generate a year's worth of electricity.

In 1975 Gordon Moore predicted that integrated circuit performance would double every two years. What has come to be known as Moore's Law has proven so accurate that chipmakers use it for long-term planning. Something similar is happening to solar cells. In 1956 cells cost $300 per watt, prices dropped to $50/W in the 1970s, and currently the price hovers around $0.85/W. According to the International Energy Agency (IEA), solar power will be price competitive with coal and nuclear generation within 10 years. The IEA predicts that by mid-century, nearly a quarter of the world's electricity will be provided by the sun.

Critics sometimes say that large-scale solar power is practical only in deserts, which are far from consumers, or in even more distant outer space. Yet the Australian Antarctic Division has installed solar panels on Macquarie Island. The 24-hour solar day compensates for the six months of winter darkness. And Germany, cold and wet for much of the year, is the world's leader in solar energy production with some 1.2 million rooftop collectors installed. In the first six months of 2012, these collectors diverted 14.7 terawatt-hours (TW-h) of surplus power to the national grid. One TW-h equals one trillion watt-hours (W-h) or enough electricity to supply Vermont for two months.

It is also said that solar cells remain too expensive to be more than a middle-class indulgence. Yet Bangladesh, one of the most impoverished countries in the world without so much as a grid connection for three-quarters of its population, will have a million solar collectors by 2015.

Grameen Shakti ("Village Energy"), an NGO founded by Bangladesh's largest bank, is the force behind this remarkable development. Shakti works out of 1500 village centers. These centers provide microloans for the purchase of solar collectors, customer training, and technical support. Unlike many other NGOs, Shakti works from the bottom up to promote community involvement and to respond to problems as they develop. Because women often have the most to gain from electrification and are the best candidates for microloans, women oversee operations at the local centers. Women technicians repair and refurbish solar units in the field and are heavily involved in manufacturing. Shakti currently operates 45 solar-component manufacturing plants in rural districts.

Most of the solar units are quite small, generating between 30 W and 100 W. But that is enough power to recharge cell phones and to free villagers from reliance on kerosene lamps. These units often pay for themselves with money saved on kerosene.

The arrival of light has dramatic effects. Stores and medical clinics remain open at night, and students study longer. Solar systems have increased the returns on poultry farming and rice processing. Home-based solar has also been responsible for a resurgence of handicrafts, which were almost forgotten arts prior to the introduction of this technology, and the establishment of night schools, which make adult education possible.

Crystals of silicon—the stuff of ordinary beach sand—has the remarkable propensity to shed electrons when exposed to light. In 1921 Albert Einstein received a Nobel Prize for his explanation of how the photovoltaic (PV) process works. The original technology for growing silicon crystals begins by melting highly purified silicon in a rotating retort. A small seed crystal lowered into the melt triggers the process. At the correct temperature and with the correct combination of rotation and extraction speed, a cylindrical crystal forms around the seed. A diamond saw slices the crystal into 200-micron (μm) thick wafers, which are then mounted on a base and given a protective covering of tempered glass. Voila! We have a highly efficient, but expensive, monocrystalline solar cell. These cells are used in applications where space restraints and durability requirements justify the higher costs. Like diamonds, monocrystalline silicon should last forever.

Polycrystalline cells came on line in 1981. They are made by pouring melted silicon into a rectangular mold and slicing it into wafers. The process saves material, but produces a complex mix of crystal boundaries and the loss of some electrical efficiency; however, the lower costs of the process more than compensate. The boundaries also refract light, which accounts for the gem-like iridescence of polycrystalline cells. Until recently, virtually all consumer-level PV was polycrystalline, although the details of the manufacturing process vary among manufacturers.

Solar panels assembled from polycrystalline silicon cells convert from 12 percent to 14 percent of impacted light energy into electricity. Each of these cells, regardless of its area, produces about 0.6 V. Amperage depends upon area: a 4 in.^{2} cell generates 3 A in bright sunlight. And contrary to what one might suppose, these and other PV cells work best in cold weather.

Thin-film technology is the term for depositing silicon or (very rarely) cadmium-telluride directly on a glass substrate. While the efficiency of thin film is nothing to brag about, the economics are favorable enough to permit the construction of large, high-voltage arrays, which are sometimes used to power submersible well pumps. Be aware that thin-film silicon loses about a fifth of its output during the first six months of sun exposure. In addition, cadmium-telluride, because of cadmium's toxicity, presents a disposal problem. It can be expensive to get rid of these panels 25 years or so down the line when they go dead.

PV technology, driven by entrepreneurs and state-supported research centers, remains in a state of flux, with new developments almost weekly. For example, San Jose-based Twin Creeks Technologies recently announced that it is ready to ship 20-μm silicon wafers, a tenth of the thickness of sawn wafers. This is accomplished by introducing hydrogen molecules at a precise depth into the base material. Heating expands the hydrogen and the brittle silicon splits, like a tree struck by lightning.

As shown in Figure 6-1, solar panels are modular constructs. Cells combine to form modules and modules combine to form panels, which are sealed against the weather with tempered glass and framed. Standard panels have a rating of 12 V. When larger amounts of power are needed, panels are grouped into arrays. Most residential systems consist of a two-panel 24 V array or a four-panel 48 V array.

Cooperating with the Sun

Solar radiation is a gift over which we have little control. Intensity varies with the latitude, season, time of day, and local weather conditions (Figs. 6-2 and 6-3). The sun remains low on the horizon at high latitudes. As we move closer to the

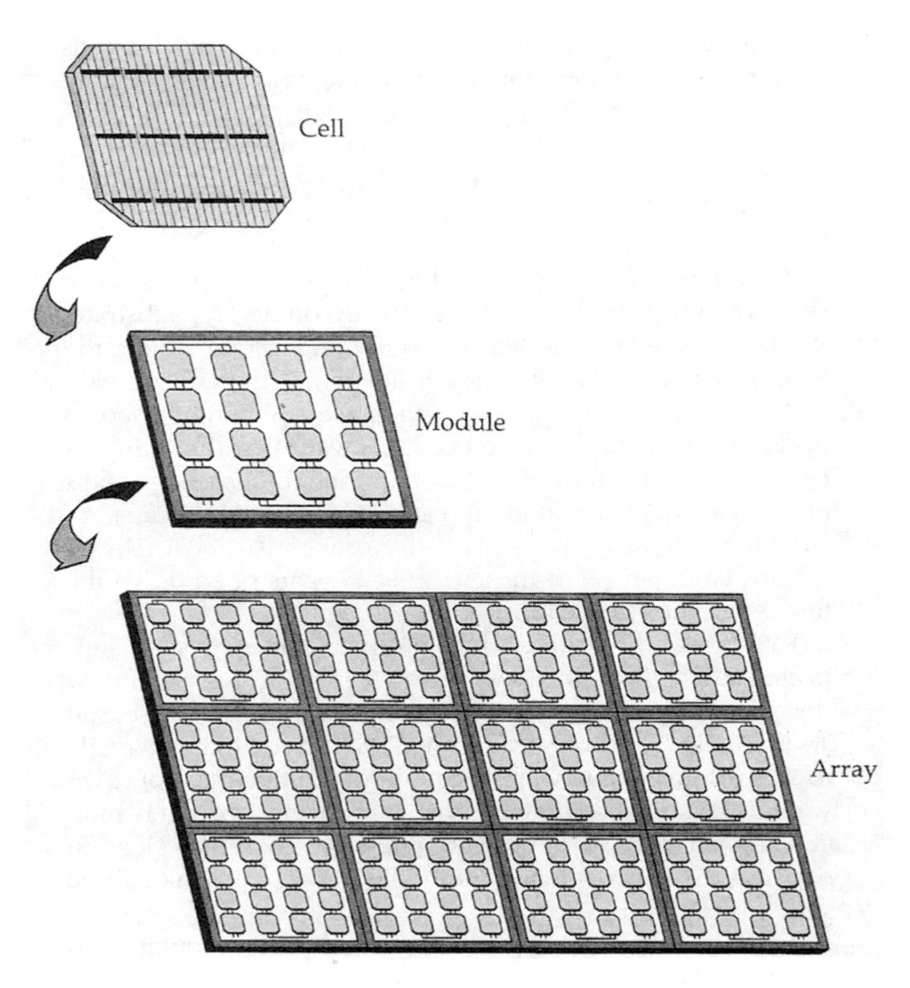

Figure 6-1. *PV cells combine into modules, and modules make up a solar panel or, as labeled here, an array.* NASA

equator, the sun climbs higher in the sky and its rays are less diffused by the atmosphere. In addition, the elliptical rotation of the Earth around the sun reduces the amount of available radiation during winter months, which is further reduced by the 23.5 degree tilt of the Earth to its axis of rotation. The tilt makes the days longer in the northern hemisphere from the spring (vernal) equinox to the fall (autumnal) equinox, that

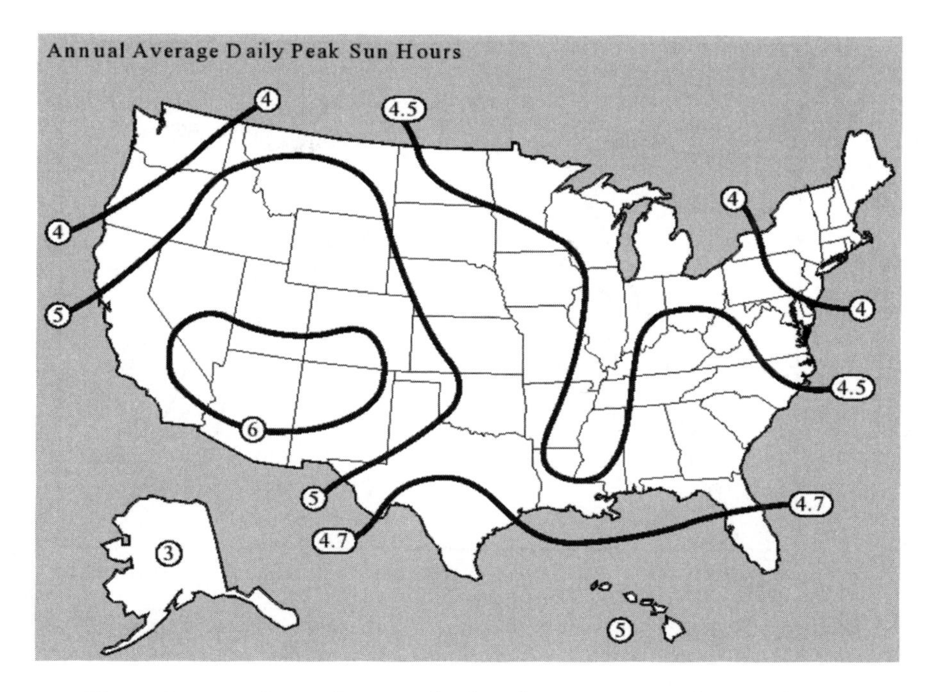

Figure 6-2. *Peak sun hours calculated as 1000 Watt-hours of radiation per square meter.* Department of Energy

is, from around March 23 to September 22. The situation is reversed in the southern hemisphere.

The National Renewable Energy Laboratory (NREL) has published solar insolation data for 239 stations located across the United States and in several of its possessions. Interested readers can access this data at http://www.nrel.gov/docs/legosti/old/5607.pdf or by calling the NREL Technical Inquiry Service at 303-275-4099. Ask for a copy of the *Solar Radiation Data Manual for Flat-Plate and Concentrating Collectors* (Task PV360501).

In North America, solar panels point due south and, as a baseline adjustment, are tilted to correspond with latitude. For example, a panel in Denver, which is near 40 degrees latitude, will function year round if tilted 40 degrees off the horizontal. For better performance, increase the tilt 15 degrees more

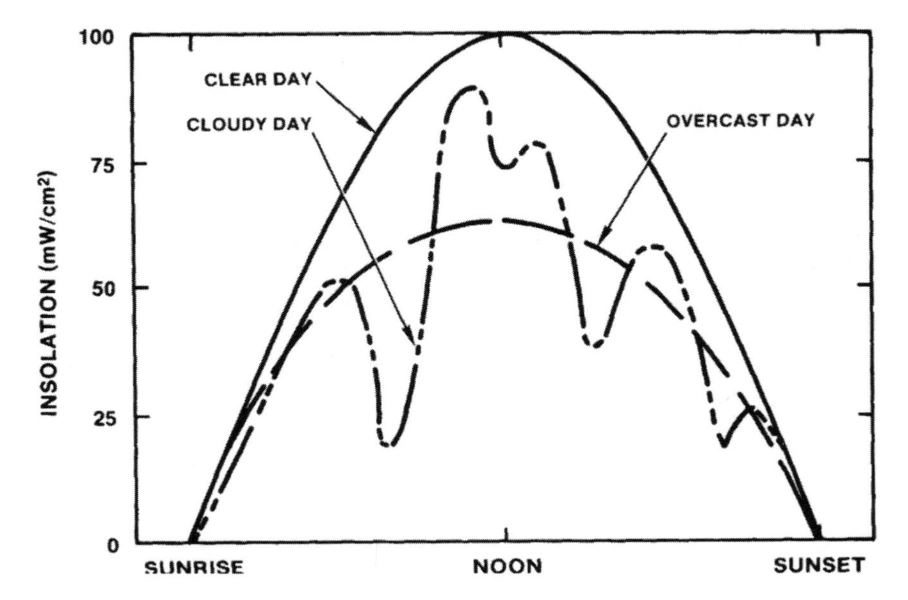

Figure 6-3. *Insolation—the amount of radiation striking a solar panel—varies during the day and with the weather. Fortunately, PV cells respond to scattered light and continue to function when the sun is obscured by clouds.* Sandia National Laboratories

than latitude in winter and 15 degrees less in summer (Table 6-1). This semi-annual vernal adjustment boosts power output in central U.S. locations by about 8 percent in summer and 5 percent in winter. A two-axis tracking system, like the one shown in Figure 6-4, makes these adjustments automatically.

Table 6-1
Panel Orientation

	Latitude	Summer	Winter	Average
Lower Canada	50°	35°	65°	50°
Upper U.S.	45°	30°	60°	45°
Middle U.S.	40°	25°	55°	40°
Lower U.S.	35°	20°	50°	35°

Figure 6-4. *A Sadoun two-axis tracking system with four 60W solar panels. According to the manufacturer, the system can boost output by as much as 35 percent. It also includes an anemometer that signals the stepper motors to lay the panels flat if wind speed becomes excessive.*

Install panels in an unshaded area mounted high enough to be clear of rain splash and snow. Shaded PV cells not only produce little or no power, they act as resistors to reduce the output of the entire array.

Sizing the System

Manufacturers determine panel output at 25°C (77°F) and under 1000 W/m^2 of solar radiation. Table 6-2 lists specifications for one popular 24 V panel.

Pmax—peak power. The most important variable.

Voc—open-circuit voltage. No-load voltage under full sun.

Isc—short-circuit current. As load increases, voltage drops and current rises to the short-circuit level. This specification determines the DC cable gauge.

Imp—maximum current output.

W/m^2—watts per square meter. An indication of panel efficiency.

The basic sizing calculation is load watt/hours divided by peak solar hours times a fudge factor, usually 1.2. But the devil lurks in the details. Consult with the vendor to arrive at a reasonable power estimate, based on pump power draw, total well depth, gpd of demand and cable length.

Controllers

DC pumps can run directly off one or more panels, an approach that has the virtue of simplicity. However, the addition of a controller improves system performance with a linear current booster (LCB) that keeps the pump running during low-light levels. Other features include a panel disconnect switch and a

Table 6-2
Panel Specifications

Model	Mfg.	Pmax (W)	Voc	Isc	Imp	Size (mm)	W/m^2	Weight (lb)	Type
GEPVp-205-M	GE Energy	205	33	8.2	7.6	981 × 1485	140	39	polycrystalline

connection for a storage-tank float switch. "Smart" controllers, sold in a package with the pump, also provide the following:

- Soft starts achieved by progressively increasing electrical power
- Protection against motor stalls and dry running
- Data on motor rpm and pump output
- Integration of diesel or wind-powered generators
- Some diagnostic capability.

Battery Banks

A battery backup keeps the pump working 24 hours a day. However, lead-acid batteries are poor repositories of electrical energy, returning far less power than required to charge them. And unless you invest in deep-discharge batteries of the kind used for industrial trucks, battery life can be measured in months. Regardless of the type of batteries used, a charge controller is essential.

A way to minimize battery headaches and still have water at night is to include a large storage tank in the system with a small booster pump to keep the pressure tank filled (Fig. 6-5). The pump can be connected to the grid or powered from a small battery bank.

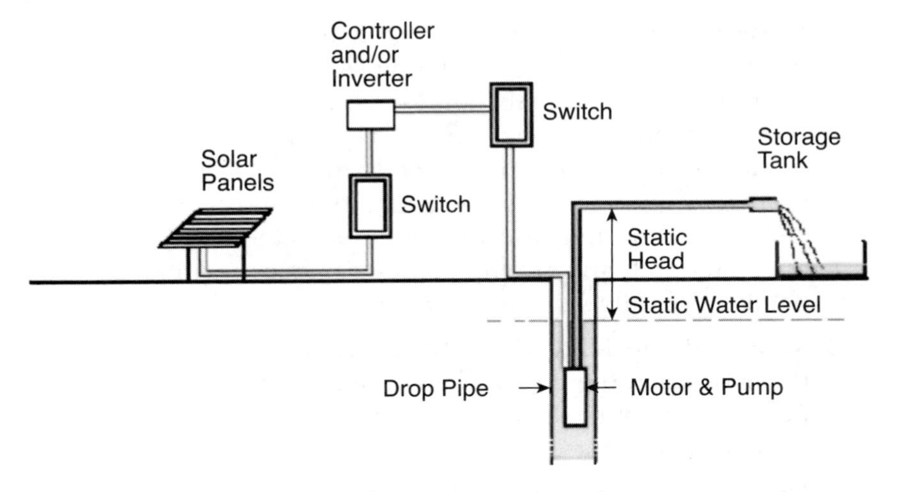

Figure 6-5. *A standby generator or a storage tank eliminates the need for a large battery bank. This figure is not drawn to scale.* U.S. Agricultural Research Service

Inverters

An inverter steps up low-voltage DC to the 115 V AC or 230 V AC needed to power AC subs and conventional home appliances. The inverter must be sized for the load and capable of providing the surge current necessary to start the sub motor, a task that can require six or more times normal amperage. It should also deliver AC in the form of a sine wave with less that 3 percent harmonic distortion. If your budget permits, avoid inverters that generate what their makers describe as a "modified sine wave." The output is actually a square wave that reduces transformer and motor efficiency by 20 percent and is unsuitable for computers.

Surface Pumps

AC pumps deliver large amounts of water in short bursts. In comparison, solar pumps produce a trickle of water, but run so long as the sun shines. Small amounts of water add up. Six gpm amounts to 360 gallons per hour. Table 6-3 provides an overview of the capabilities of the various types of solar surface pumps.

There are four things to consider when shopping for a pump:

1. **Power requirements** Pump voltage and power draw (watts = volts × amperes) determines how many arrays are required and how the arrays are interconnected. Two 50 W arrays or one 100 W array satisfy the power requirements for Shurflo 2088 and other small surface pumps. As pump wattage goes up, more arrays are needed. Connecting arrays in series adds the voltage produced by each array to the output. Connecting arrays in parallel does the same for amperage.
2. **Suction lift capability** How much suction lift you need depends upon the depth of the water level below the pump. None of these pumps will raise water from wells deeper than about 25 ft and most do less.
3. **Dynamic versus positive displacement** Centrifugal and centrifugal jet pumps—dynamic pumps—are the best at

Table 6-3
Representative Solar Surface Pumps

Mfr. & Model	Output	Suction Lift	Total Dynamic Head	Voltage	Watts	Type
SunPumps SJT	2–200 gpm	20 ft	50 ft	60, 75, 90 V DC	200–550 W	Jet
Dankoff SunCentric	5.0 to 70.0 gpm	10 ft	90 ft	12, 24, 36, 48 V DC	168 W	Centrifugal
Shurflo 2088	Up to 3.5 gpm	12 ft	104 ft (45 psi)	12, 24 V DC, 115 V AC	84 W	Diaphragm, Recreational Vehicle
Dankoff Slowpump 1400 and 2600	0.5 to 4.0 gpm	20 ft	Up to 440 ft	12, 24, 48 V DC or 36 V DC direct coupled to solar panel, 115 V AC @ 0.9 A	451 W max.	Rotary vane
Dankoff Solar Force	5.0 to 9.0 gpm	25 ft	Up to 230 ft	12, 24, 48 V DC, 115 or 230 V AC	n/a	Piston

moving water so long as they run at their rated speeds. At half speed, output drops by 75 percent or more.

Positive-displacement pumps, a category that includes piston, vane, and diaphragm types, deliver the same volume of water during each revolution. So long as the armature turns, some water will be pumped. The advantage for solar applications is obvious.

Piston pumps, with their heavy cast-iron frames, brass cylinder barrels and leather piston seals are the most durable positive-displacement pumps and, if the need arises, can easily be hooked up to a small engine. Recip–rocating machinery has a kind of elemental appeal, at least for those of us who cut our teeth on connecting rods and crankshafts.

4. **Cost versus Durability** The trade-off between initial cost and durability is severe. You can spend thousands for a beautifully engineered Dankoff piston pump that will soldier on for decades with no more than a brush replacement, or spend $40 for a diaphragm or flexible-vane pump. However, these inexpensive pumps, intended for recreational vehicles, fail quickly in regular service.

Submersible Pumps

Most water wells are deep enough to require a submersible pump.

Diaphragm

Diaphragm pumps are a good choice for moderately deep, low-yield wells (Table 6-4). These positive-displacement pumps require little energy and can be suspended on ½-in. PVC or poly drop pipe, which makes the string light enough for one person to retrieve.

The diaphragm is a sacrificial item. One manufacturer suggests that it be replaced every 24 months, but while some fail early, others survive for years without problem. Repair costs range from around $100 for a diaphragm and its related parts to $300 or so if a dealer does the reconditioning.

Table 6-4
Diaphragm-type submersible pumps

Mfr. and Model	Output at Max. depth	Max. Depth	Voltage	Type
Shurflo 9300	1 gpm	230 ft	24 V DC	High head
Sun SDS-D-228	0.82 gpm	230 ft	12–30 V DC	High head
Sun SDS-Q-128	2.70 gpm	100 ft	12–30 V DC	Quad
Robinson BL40Q	2.0 gpm	185 ft	12.6–40 V DC	Quad

When you need lots of water, there is no substitute for a centrifugal pump, although power requirements go up dramatically when high outputs from deep wells are needed. Helical pumps, being positive-displacement devices, deliver less water but develop more head pressure (Table 6-5). The helical rotor, which looks like a stretched coil spring, turns inside an elastomer stator, consisting of a series of twisted cavities (Fig. 6-6). Helical pumps have become popular for stock watering from depths of 70 ft to 150 ft. Table 6-5 compares the performance of smaller Lorentz helical and centrifugal pumps.

DC versus AC

AC subs are the default choice for many owners. Having been in production for decades, AC subs cost less than the newer DC types and are familiar to service personnel. Some drillers will have nothing to do with DC machinery. However, AC power requires an inverter and either a high-voltage solar array or a step-up transformer, equipment that does not come

Table 6-5
Smaller Lorentz helical rotor (suffix "HR") and centrifugal ("C") pump data. The company also makes larger pumps in both configurations.

	PS 200HR	PS 600HR	PS 150C	PS600C
Max. total lift	170 ft	590 ft	65 ft	80 ft
Flow	700 gph	700 gph	1300 gph	2900 gph
Nominal voltage	24-48 V DC	48-72 V DC	12-24 V DC	48-72 V DC

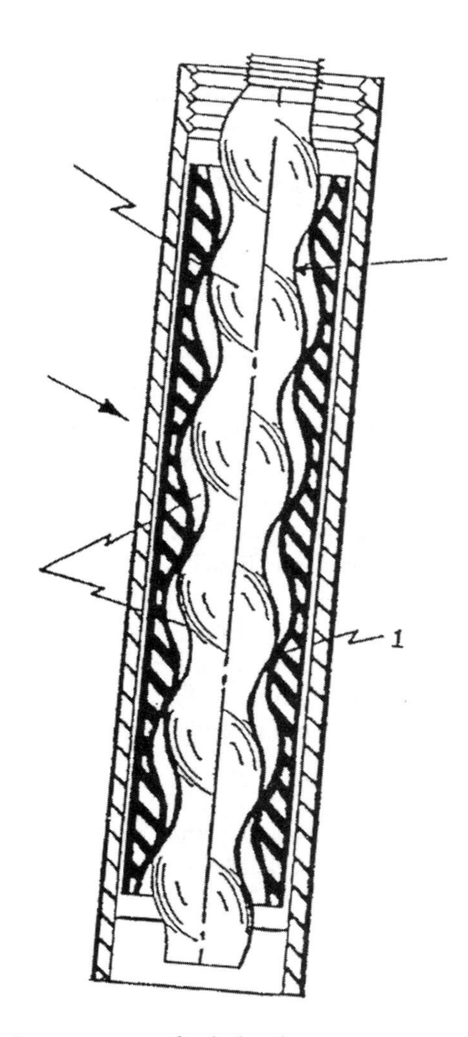

Figure 6-6. *Cross-section of a helical, or progressive cavity, pump. The elastomer stator, usually made of Buna-N, consists of a double thread. The rotor turns eccentrically, moving fluid through a series of cavities it forms with the stator.* U.S. Patent No. 6354824B1

cheaply. The economics look better when long cable runs between the solar array and pump are involved. Since electrical resistance is a function of amperage, high-voltage, low-amperage AC enables lighter gauge cable to be used.

Several manufacturers offer brushless, direct-current (BLDC) motors for solar applications. A BLDC motor employs permanent magnets in its armature that react with the stator coils to generate torque. Solid-state electronics function like brushes to create a rotating magnetic field around the armature. The circuitry also eliminates the heavy surge of current AC motors draw during starting. BLDC motors consume less current than brushed motors for the same output and, consequently, can be made smaller. And elimination of the mechanical/electrical interface created by brushes should do wonders for reliability.

On the other hand, the control circuitry is complex and the rare-earth armature magnets are expensive. BLDC subs cost more than their AC equivalents, and service personnel, some of them anyway, struggle with the learning curve.

Wind Power

At the end of the nineteenth century, Western Europe was home to 200,000 windmills, most of them in Spain, Portugal, and Scandinavian countries where water power was scarce. Windmills ground paper pulp, sawed wood, polished glass, and pulverized chalk for cement. In the Netherlands, windmill-driven, screw-type pumps reclaimed land from the North Sea. Even as late as 1900, windmills ground the entire wheat crop of northern Europe.

One of the most famous windmills—the South Murphy—has recently undergone a complete restoration as tourist attraction. Constructed in 1907 to pump water in San Francisco's Golden Gate Park, the South Murphy stands 95 ft high and has 114-ft sails. During its brief operational life, the mill pumped as much as 40,000 gallons per day.

Direct-Drive Windmill Pumps

As shown in Figure 6-7, the classic farm and ranch wind pump mounts on a steel tower, or mast, with the water discharge going to a nearby stock tank. A gearbox/eccentric/pump rod mechanism drives a downhole pump. The rotor wheel comes

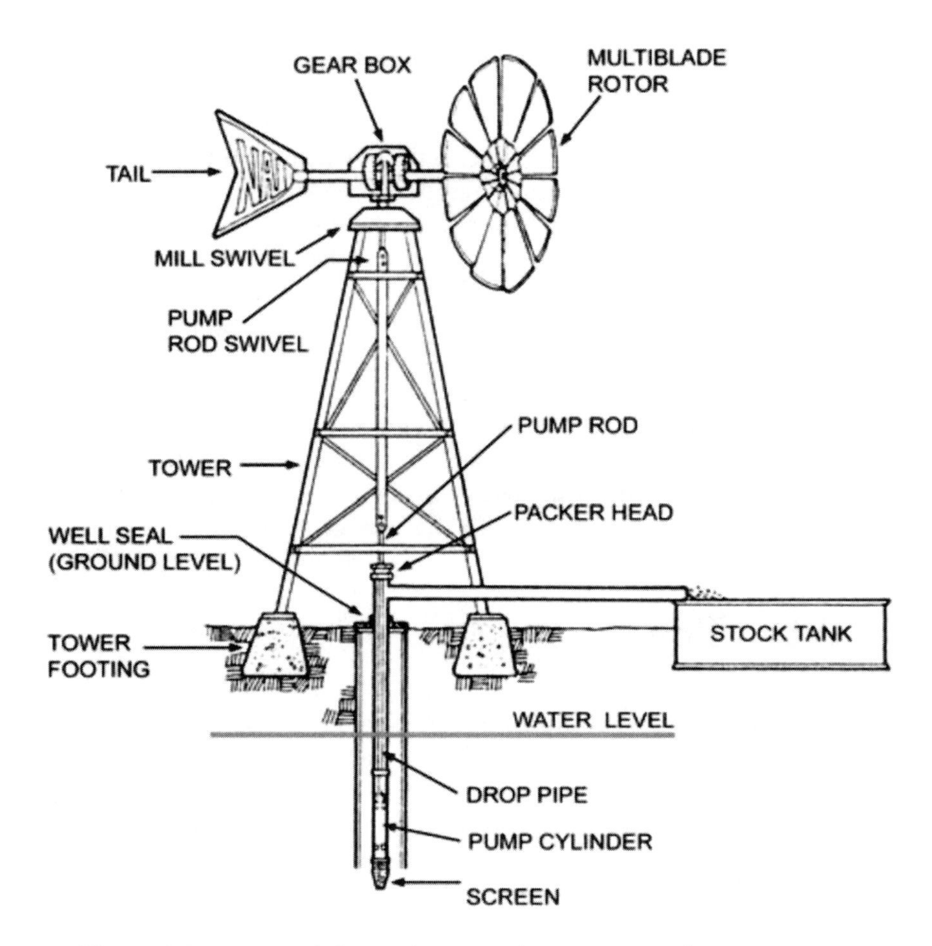

Figure 6-7. *A typical direct-drive wind pump set up for watering stock. See Fig. 6.8 for detail on the pump cylinder.* British Columbia, Ministry of Agriculture and Lands, Stock Watering Factsheet

in diameters of 6 to 20 ft, with 8 ft being the most popular. Some wheels have as many as 40 sheet-steel blades. Although these pumps look archaic, the design is quite sophisticated, evolved by trial and error through centuries of experience. The almost solid front presented by the rotor means that a barely perceptible breeze gets the blades turning.

Performance peaks at wind velocities of around 15 mph. At higher wind speeds, drag and turbulence come to the fore, and efficiency, which is never very high, drops. The off-center pivot bearing and oversized tail vane cause the rotor to furl—to tuck itself alongside the tail boom—at a preset wind speed.

The power output is nothing to brag about (Table 6-6), but large wind pumps can lift water from 1000-ft wells.

Pumping takes place on the upstroke, which means that the rotor wheel must develop enough torque to lift the weight of water in the drop pipe and the weight of the sucker rod (Fig. 6-8). As one might suspect, this can inhibit starting. Fiberglass or floating pump rods alleviate the situation somewhat.

Newer, counterbalanced pumps use weights or springs to offset half of the water weight and some fraction of the rod weight. It is claimed that these pumps deliver twice the water of older models at wind speeds below 10 mph and a third more water at higher speeds. Counterbalancing also reduces stress on the sucker rod.

Another approach is to operate the pump through a cam, rather than an eccentric. The cam removes the symmetry from the action; rather than devoting half a revolution to lift, the cam spreads out the load to three-quarters of a revolution. Cam-type pumps are said to lift 70 percent more water than equivalent eccentric pumps during low-wind conditions.

Table 6-6
Windmill power output during a 15-mph (24-kph) wind, assuming 20% efficiency

	Rotor Diameter					
	6 ft (1.8 m)	**8 ft (2.4 m)**	**10 ft (3 m)**	**12 ft (3.7 m)**	**14 ft (4.3 m)**	**16 ft (4.9 m)**
Horsepower	0.12	0.21	0.33	0.50	0.70	0.85
(Watts)	(90)	(160)	(250)	(380)	(510)	(640)

Source: British Columbia, Ministry of Agriculture and Lands, Stock Watering Factsheet

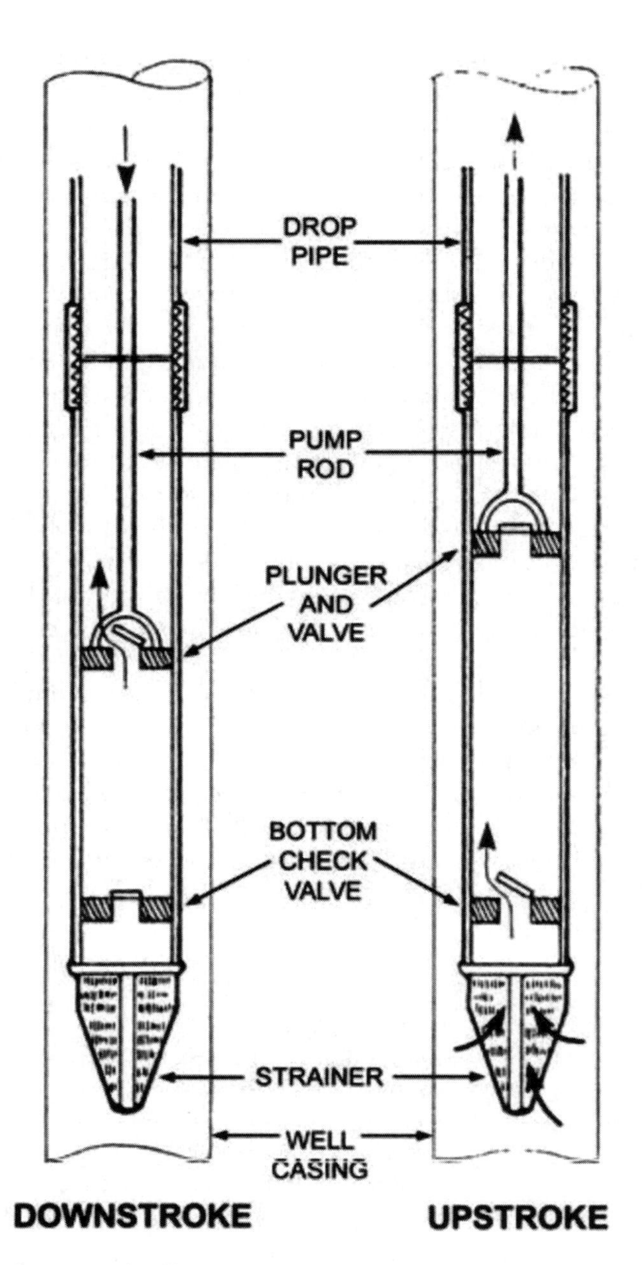

Figure 6-8. *Windmill pumps vary in detail, but all incorporate inlet and outlet check valves in the piston. Modern examples use urethane, rather than leather, seals.* British Columbia, Ministry of Agriculture and Lands, Stock Watering Factsheet

Sizing

Determine the vertical lift by measuring from the surface to the water level in the well and adding any additional lift required to fill the water tank. Manufacturers describe the performance of their products at maximum wind velocities, which can be as high as 30 mph. Under anything approaching real-world conditions, these velocities may only occur a few hours a day.

In addition to wind velocity, wheel diameter and pump cylinder diameter and stroke determine output. The stroke is adjustable. Most manufacturers, following the Aermotor example, have arranged matters so that the short stroke option increases lift by a third and reduces gph output by a quarter. Cylinder diameters range from 1¾ in. to 7 in. While it's difficult to generalize, a 1¾-in. pump, driven by an 8-ft wheel at a wind speed of 15 mph should deliver 2 gpm at a lift of 175 ft. Under the same conditions, a 3-in. pump might raise 5 gpm at 70 ft. The manufacturer's literature will be more precise.

Siting

Because the long pump rod must be concentric with the casing, wind pumps sit directly over the well. And because the windmill pivots with changing winds, the rotor wheel must be dead level if it is to respond correctly. Old timers used telescopic rifle sights to center the pump rod.

Difficulties arise in forested or built-up areas. In order to avoid turbulence, the rotor wheel should be elevated at least 30 ft above any obstructions within a 400-ft radius. The fewer the obstructions and the higher the wheel, the better.

Maintenance

While I cannot speak for the imports, domestic wind pumps are rugged machines, some of which are approaching a century of service. But maintenance is necessary. Every six months make a careful visual inspection of the structure and the blades. Tighten the fasteners and be alert for any unusual noises. Change the gearbox oil every year. Most require 10-W, nondetergent oil, which can be ordered from Aermotor. Pump seals should last five or six years.

Rebuilding

Rebuilding wind pumps has become a cottage industry, with small one- or two-men shops scattered throughout the farm and ranch belt. Vendors, such as WindMill-Parts (940-597-7735, windmill.parts@juno.com), carry refurbished pumps and parts, including square-headed bolts to make the job look authentic.

Try to find an Aermotor (800-854-1656, http://aermotorwindmill.com) of late enough vintage to have an oil-bath gearbox (Fig. 6-9). A Dempster (800-777-0212, sales@dempsterllc.com) is

Figure 6-9. *A vintage Aermotor, located in Springview, Nebraska.* Courtesy of DOE/NREL; Photo credit: Warren Gretz

another good prospect. Both companies can supply parts for their old machines. Parts that fit Aermotors can also be had from Dean Bennett and other vendors of Fiasa wind pumps. Several years ago Aermotor farmed out its foundry work to Fiasa, an Argentine manufacturer of agricultural machinery. The arrangement was short-lived, but it enabled Fiasa to go into production with an Aermotor 702 clone. Parts are almost 100 percent interchangeable with the American product.

Rebuilding a wind pump is a serious project. You will need a crane truck and a crew of experienced riggers to dismount and mount the unit, which can weigh upwards of 400 lb. Gears will be worn and shafts out of round. Pre-1940s pumps had Babbitt bearings, which must be poured in place, with the molten metal contained by clay dams. Pouring Babbitt is not rocket science, but you will need to research the subject—*How I Pour Babbitt Bearings*, by Vincent Gingery is a good basic text—and make a few practice pours to develop a feel for the work.

HAWT Generators

Wind turbines are classed by the lay of the hub shaft. VAWTs (Vertical Axis Wind Turbines) enjoy passionate support among a minority of home builders, but have achieved only limited popularity because of structural and, one suspects, esthetic issues. The discussion that follows focuses on conventional, horizontal-axis (HAWT) machines.

Although wind pumps and turbine generators both lift water, they represent different philosophies. Wind pumps harken back to the grain mills from which they evolved. These slow-turning, heavy, and less-than-precise machines run for years with minimal attention. And when something goes wrong, repairs tend to be simple.

HAWTs are lightweight aerodynamic constructs that grew out of research on propeller-driven aircraft. Failure is usually catastrophic, and repairs are far from simple. Efficiency, which can be twice that of a wind pump, comes at the cost of complexity. On the other hand, wind pumps merely lift water. The electricity generated by wind turbines has a multitude of uses.

Table 6-7 lists specifications for turbine generators with nameplate ratings of 900 W to 11 kW.

Table 6-7
Representative HAWTs

Mfr.	Model	Rated Output	Rotor Diameter	Voltage	Rated Power@ Wind Speed	Cut-In Speed	Blades
Southwest Windpower	Whisper 100	900W	7.0 (2.1 m)	12, 24, 36, 48 V DC	28 mph (12.5 m/s)	7.5 mph (3.4 m/s)	3 carbon/ fiberglass
Ampair	600	1050 W @24 V DC, 1140 W @ 48 V DC	5.6 ft (1.7 m)	24 or 48 V DC	24.6 mph (11.0 m/sec	17.9 mph (8 m/s)	3 GPR
Bergey	BWC 1500	1500 W	10 ft (3.0 m)	12–120 V DC	28 mph (12.5 m/s	8 mph (3.6 m/s)	3 unknown composite
Southwest Windpower	Skystream	2400 W	12 ft (3.7 m)	120/240V AC, 120/208 V AC	29 mph (13 m/s)	8.0 mph (3.6 m/s)	3 fiberglass/ composite
Ampair	6000 MWI	6000 W 4800 W to battery	18 ft (5.5 m)	48 V DC or 115/230 V AC	24.6 mph (11.0 m/s)	7.8 mph (3.5 m/s	3 Glass-filled polypropline
ARI Green Energy	ARI7.5	7500 W	21 ft (6.4 m)	240 V DC	22.4 mph (10 m/s)	5.6 mph (2.5 m/s)	3 fiberglass/ composite
Bergey	BWC Excell	10,000 W	23 ft (7.0 m)	3-phase 240 V AC, 48 to 240 V DC	31 mph (13.8 mps)	7.5 mph (3.4 mps)	3 fiber-reinforced plastic
Gaia-Wind	11 kW	11,000 W	42.7 ft (13 m)	AC grid matched, 460 V AC	21.3 mph (9.5 m/s	5.6 mph (2.5 m/s)	2 epoxy/ fiberglass

There are two facts to remember about turbine generators:

1. Power output is proportional to the cube of wind speed. Double the wind speed, and the power output increases eight times.
2. Power output is directly proportional to the swept area of the rotor. Double the swept area, and power output doubles.

Siting

At least an acre of ground that receives average annual wind velocities of more than 12 mph is required to site a wind turbine (Fig. 6-10). Small turbines need 5 to 8 mph winds to generate any power at all.

You may want to download the *National Wind Energy Resource Atlas of United States,* at www.windpoweringamerica.gov. Several states publish county-by-county wind profiles that vary seasonally, with most wind energy available in winter.

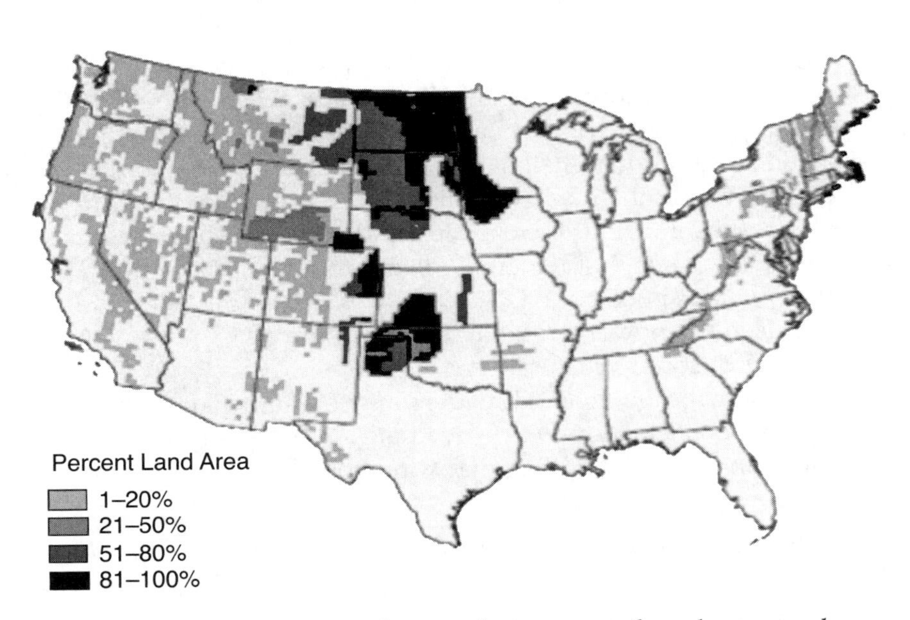

Figure 6-10. *A DOE map showing the presence of yearly averaged Class 4 (12.5–13.4 mph) winds in the continental U.S.*

Another excellent source is www.windfinder.com, which reports wind histories from thousands of stations, worldwide.

The surest way to evaluate a site is with a recording anemometer mounted at turbine hub height (Fig. 6-11). Data collected will show average wind velocity and, what's almost equally important, the distribution of velocities. Local wind-power enthusiasts sometimes loan the equipment.

Do not expect anything like steady power from a wind generator. Even the best mountaintop and desert sites do not exceed 40 percent availability. From the scattered data available, it appears that U.S. wind farms function about 30 percent of the time. Conventional generating plants make up the shortfall. Small turbines should be backed up with solar, battery, or grid power. In some applications, batteries are considered the primary source of power, with wind energy confined to recharging.

Towers

Ideally the tower should be at the center of a cleared circle, with no trees, buildings, or other ground clutter closer than 350 ft. In any event, the lower edge of the turbine blades should be clear of the turbulence obstacles generate (Fig. 6-12)

Elevating the tower has another advantage. The phenomenon known as "wind shear" means that horizontal wind speed increases with height. For example, if our turbine sees 20-mph winds at 25 feet, doubling the height of the tower to 50 ft results in a 1.10 increase in wind velocity. That doesn't sound like much, but remember that turbine output is a function of wind speed cubed. $1.10 \times 1.10 \times 1.10 = 1.33$. We have increased the power potential of our turbine by 33 percent at the cost of some structural steel.

Mathematically inclined readers might want to know how the above calculation was arrived at. It's based on the "1/7 Rule," an empirical finding that wind speeds increase by the 1/7 power of the height.

25 ^ 1/7 = 1.58, or the 7th root of 25.

50 ^ 1/7 = 1.74, the 7th root of 50

$1.74 \div 1.58 = 1.10$, the gain in wind velocity

$1.10 \times 1.10 \times 1.10 = 1.33$, the boost in turbine output.

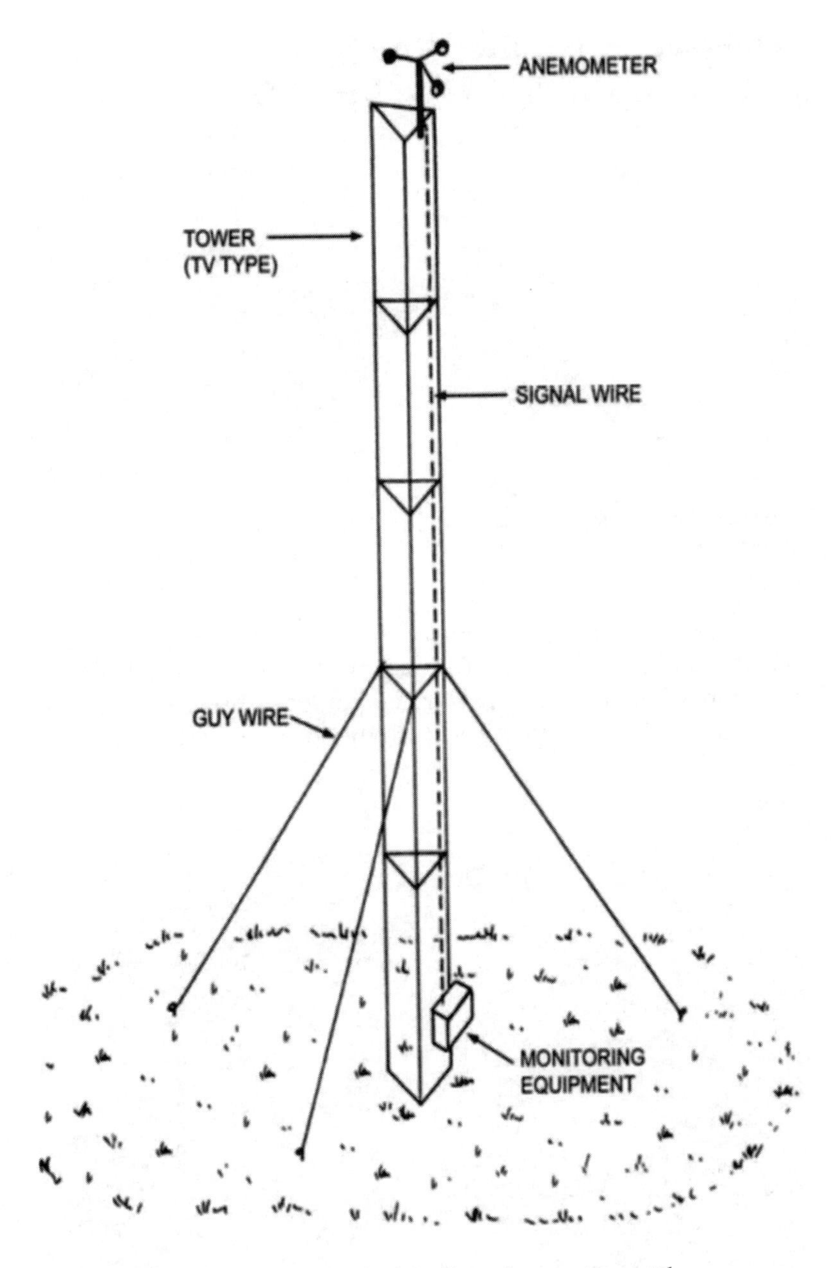

Figure 6-11. *Wind velocity data collected over time with an anemometer is the surest guide to site selection*. British Columbia, Ministry of Agriculture and Lands, Stock Watering Factsheet

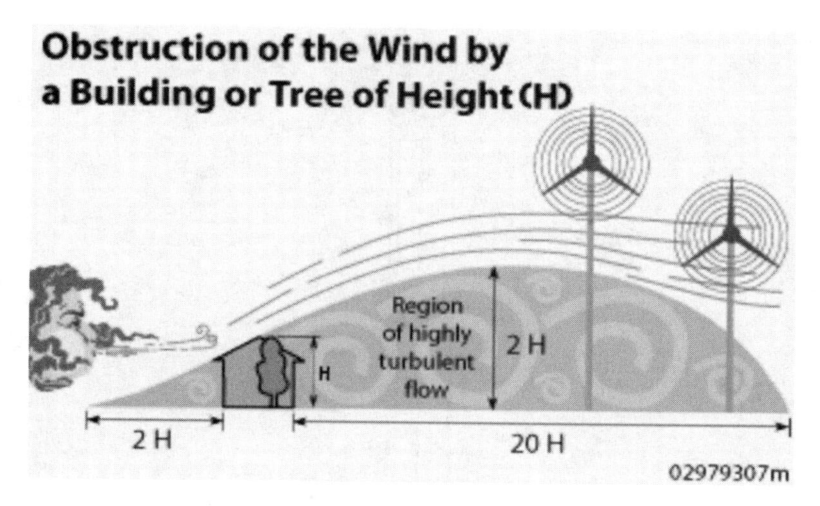

Figure 6-12. *Buildings and trees generate turbulence that compromises rotor efficiency.* U.S. Department of Energy

Concrete or lattice towers support large turbine generators; home builders usually prefer monopole towers, stabilized by guy wires (Fig. 6-13). You will need at least three steel cables, spaced 120 degrees apart and with the mooring radius one-half

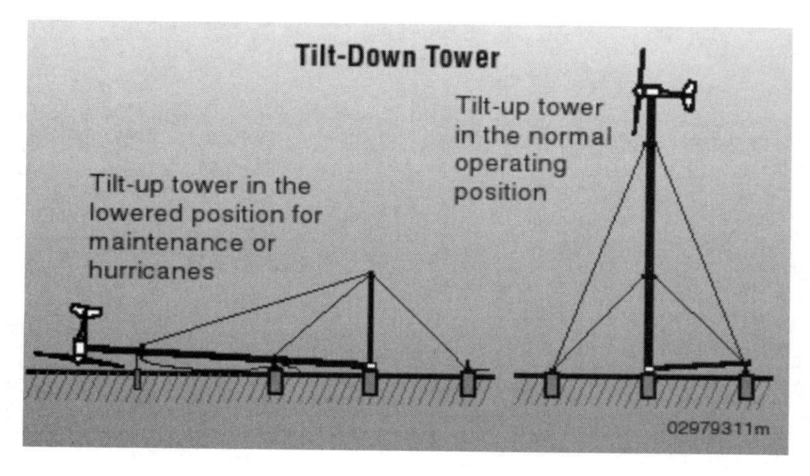

Figure 6-13. *A tilt-down tower simplifies maintenance.* U.S. Department of Energy

to three-quarters of the tower height. In other words, if the tower is 40 ft tall, the cables should be tied down in a circle with a radius of 20 to 30 ft.

Rotor Diameter

Output has a one-to-one relationship with increases in rotor swept area (Fig. 6-14). Since the area of a circle is radius

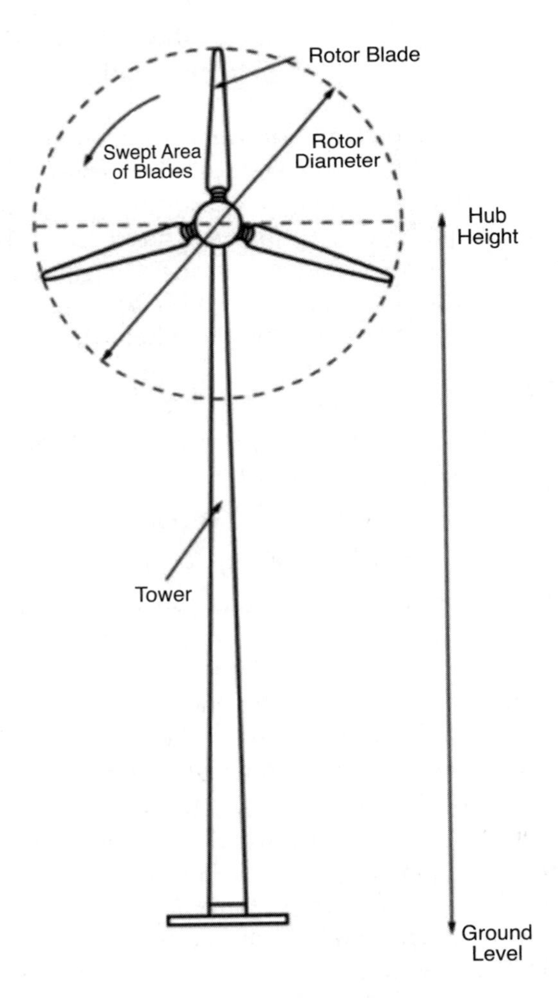

Figure 6-14. *Wind turbine nomenclature.* Government of Alberta, Agriculture and Rural Development

squared times pi ($R^2 \times 3.14$), doubling the diameter quadruples the swept area and power output. For example, a three-foot-diameter rotor transcribes an area of ($1.5 \times 1.5 \times 3.14$) 7.1 ft^2. Increasing rotor diameter to 6 ft gives us an area of 28.3 ft^2. Big rotors are the way to go.

Few customers appreciate the importance of large rotors and tall towers. Recently thousands of Europeans were inveigled into purchasing microturbines for rooftop mounting. Vendors promised that these machines, some of which produced less than 30 W, would cut their electric bills by a third. Of course, that didn't happen, and authorities are now investigating.

This is not to say that microturbines are entirely a scam. Designed and mounted correctly, these devices provide good service for customers with limited requirements. Examples include marine-quality microturbines fitted to sailboats to provide power for low-wattage radio and navigational equipment. It is also true that small quantities of electricity mean a great deal to people living in places where there is none.

Tip-Speed Ratio

Spec sheets often include the tip-speed ratio (TSR), or the ratio of tip velocity to wind velocity:

$$\text{TSR} = \text{tip velocity}/\text{wind velocity}.$$

If the tip speed is 75 mph and wind velocity 25 mph, the TSR is 3.0, or $75 \div 25$. This simple concept has important implications.

The air stream approaches the turbine head-on. As it impacts a blade, it gives up kinetic energy, which is the purpose of the exercise. But in doing so, the direction of the air stream is diverted, and flow becomes turbulent. The turbulent zone trails behind the turbine and extends for a short distance in front of it. Fast-moving water behaves the same way when it encounters an obstruction. Remove the obstruction and, a short time later, flow returns to normal

The space between blades is the equivalent of removing the obstruction. The following blade, representing a new obstruction, then comes into play. It should arrive just as flow returns to normal. If early, the blade enters the turbulent zone that gives up little kinetic energy. Turbine efficiency drops. If the blade arrives late, sometime after flow returns to normal, much of the wind energy has already escaped through the gap between the adjacent blades.

The more blades we have, the less of a gap between them and the greater the drag, or resistance to turning. A solid rotor would have a tip-speed ratio of 1.0 and the drag of a barn door. It would not turn. The multiple blades of Aermotor-style wind pumps have a TSR of around 1.6. A three-bladed rotor comes out to around 5.0; a two-bladed rotor approaches the optimum for electrical generation with a TSR of 6.0 or thereabouts. The smaller the number of blades, the higher the TSR and the faster the turbine must turn to extract energy from the wind.

Manufacturers supply tip-speed data, but if you're building your own, you'll have to make the calculation in order to avoid turbulence and match blade speed with generator characteristics. Tools required are a tape measure, digital tachometer, and digital anemometer. These instruments can be purchased quite inexpensively.

Let's assume that the blade radius—the distance from the center of the hub shaft to a blade tip—is 6 ft and the turbine turns 300 rpm at a wind speed of 23 mph.

The distance the blade describes during one revolution:

$$\text{Circumference of a circle} = 2R \times \text{pi}$$
$$\text{Tip travel/revolution} = 12 \times 3.14 = 37.68 \text{ ft}$$

At 300 rpm, the blade makes (300 × 60) 18,000 revolutions per hour.

$$\text{Tip travel feet per hour} = 18{,}000 \times 37.68 = 678{,}240 \text{ ft}$$

The next step is to convert ft/hr to mph:

$$1 \text{ mile} = 5280 \text{ ft.}$$
$$678{,}240 \div 5280 = 128.5 \text{ mph}$$

If our wind speed measurement gave us 23 mph:

$$\text{TSR} = 128.5 \div 23 = 5.59$$

Blade Profiles

Airfoil-shaped blade profiles, developed by the predecessor to NASA in the 1930s and 1940s, reduce drag and enable the higher blade speeds needed for electric power generation. However, high blade speeds do not come without cost. Blades must be strong enough to contain the centrifugal forces and the gyroscopic forces that develop as the turbine swings into the wind. Noise and blade erosion increase.

Serious students should obtain a copy of *Wind Power Plants: Fundamentals, Design, Construction and Operation,* Earthscan Publications, Ltd.,edited by R. Gasch and J. Twele. Professor Gasch is recognized as an international expert on airfoils and other aspects of turbine design. Another excellent source is Hugh Piggott's *The Wind Turbine Recipe Book,* which has become the bible of DIY turbine building. Some idea of the importance of this text can be had from the fact that it has been translated into a dozen languages, including Kiswahili, the primary language of Tanzania.

Most home builders either purchase blades or carve them from wood, often cypress. Others mold blades from fiberglass, and a few intrepid souls have welded up hollow blades from aluminum sheet. Blade design is a complex, math-intensive procedure. Fortunately www.warlock.com.au has come to the rescue with blade calculator software for two- and three-blade 0.70-m and 0.75-m (2.30-ft and 2.47-ft) radius turbines. Other sizes have a nominal cost. You may also want to obtain a $5 software package from www.windstuffnow.com, which includes electrical performance.

The fast and dirty way to proceed is to cut the blades from 8- or 6-in. diameter PVC pipe (Figs. 6-15 and 6-16). The airfoil shape, formed by the curve of the piping, is primitive, and the PVC dust somewhat toxic. We wouldn't repeat the experiment.

A large automotive radiator fan also works to the extent that it will generate enough power for cell-phone charging.

Power Coefficient

The power coefficient (C_p) is the percentage of mechanical power extracted by the turbine relative to available wind

Figure 6-15. *Blades cut from PVC pipe are infinitely easier to fabricate than those made of wood, aluminum, or fiberglass. Leading edges were rounded and trailing edges feathered. Subsequently, we experimented with various profiles. Using a heat gun, we progressively flattened the curvature and varied twist given to the blades to improve the angle of attack in stages from zero to 10 degrees.* Tony Shelby

Figure 6-16. *A pickup truck was used as a test cell to evaluate blade profiles and twists. Blade rpm was monitored as open-circuit output voltage from a permanent-magnet alternator.* Tony Shelby

energy. As was established by Benz in 1919, the maximum C_p obtainable is 59.26%. Large utility scale turbines achieve power coefficients of around 40%. The smaller units we deal with average around 30%.

Rotor Speeds

A wind generator begins to generate power at its cut-in speed and goes off-line at its cutout speed. Governed, or maximum speed, is the rpm limit, achieved by furling, feathering the blades, or braking. The danger of blade shedding through over-speeding or fatigue failure is another reason why turbines should be mounted in splendid isolation.

Purchasing

Like automobiles 100 years ago, wind turbines are an infant technology. Quality varies: some would do Ludwig Prandtl or Hugo Junkers proud; others are a waste of resources. The problem of selecting the right small turbine is compounded by the lack of regulation and policing. Unless a third party, such as Intertec or the Small Wind Turbine Certification Council (SWTCC), has tested the products, the performance data are whatever the manufacturer says they are. There is no certification for long-term durability.

For the reasons indicated, some of the more reputable manufacturers include:

- **Ampair**—a maker of microturbines primarily for sailboat applications. Ampair has been in business for decades, and the quality of its products is recognized by the marine industry.
- **ARI Green Energy**—a research-oriented Florida-based firm that has recently received grants to develop the next generation of small turbines and a wind-powered data center. Data centers are the largest consumers of electrical power in the country. ARI manufactures 1.8-, 7.5-, and 50-kW units with advanced blade technology.
- **Bergey**—in business for 30 years with thousands of satisfied customers for its 1-, 5-, and 10-kW turbines. The founder and upper management are engineers with backgrounds in theoretical and practical aerodynamics. And like good technologists, they prefer understatement.
- **Gaia-Wind**—the 11-kW turbine, developed in Denmark, considered by many to be the Lexus of the industry, has a design life of 20 years. Over 350 of these machines are in use worldwide.
- **Southwest Windpower**—a U.S. manufacturer that has shipped 170,000 small wind turbines to some 120 countries, making it the world leader in small turbine sales. Its Skystream 3.7-kW unit has had its performance certified by the SWTCC.

I hasten to add that I have no financial or other connection with these companies. None of them have been so kind as to send me a wind turbine.

Sizing

Because of the variability of wind, manufacturers describe output in annual terms. Purchasing a complete system with a rectifier (needed when the turbine produces AC power), voltage and current regulators, and a charge controller (for battery banks) simplifies matters. For well pumps, determine the current draw in watt-hrs per day or month and extrapolate to an annual average. The vendor should be able to suggest a system appropriate for local wind conditions.

A 1.8-kW unit should be enough to supply a hunting camp or other remote site. If your ambitions extend to reducing or, under ideal conditions, eliminating electricity bills for an average American home, you will need serious power in the range of 7.5 to 11 kW. Depending upon the location, some of the investment can be recouped by tax breaks and by selling surplus power to the utility.

Generator Notes

Some DIYers take the high road and scratch-build a three-phase alternator, sized for the turbine and fitted with neodymium alloy ($Nd_2Fe_{14}B$) permanent magnets (PMs). To simplify fabrication, rare-earth magnets can be retrofitted to Delco and other automotive alternators. Windstuffnow supplies kits and information for the conversion. Kubota yard tractor PM alternators put out 14 A at 12 V and cost about $100 on the aftermarket. Other builders favor PM treadmill motors, sometimes available on EBay. These motors have flywheels that can be drilled and tapped to serve as the blade hub.

We used a permanent-magnet stepper motor from an electronics surplus store (Fig. 6-17). Bring a VOM along when shopping. Spinning the shaft by hand should generate at least 1 V across the brush terminals. A large diode will also be needed to isolate the battery bank. Without the blocking diode, the batteries would power the turbine at low wind speeds. You will also need a wind-turbine controller consisting of a voltage regulator and some form of load dumping to avoid overheating the battery plates in high winds. Rather than dissipate itself across a resistor, the surplus current can be used to heat water. Alternate-energy stores carry the controllers.

Figure 6-17. *The PM motor chucked in a drill press and spinning at 500 rpm. The 14.4 V output at this speed translates to 34.7 rpm/V. If we assume that output is linear, the 13.8 V minimum required for battery charging means that the rotor should turn 480 rpm.* Tony Shelby

Human Power

Human muscle power is strangely undervalued. Except for athletes and circus performers, those who use it most are paid less. No government has reduced itself to taxing muscle power. Yet this is the power that built pyramids and cathedrals, and even today, feeds much of the world. Our bodies are our first and last resort.

We thought we would investigate the way human power can move water.

Pedal-Powered Pump

Pedaling in an upright position (not crouched over a racing bike) is the most efficient extractor of muscle power yet devised. According to several studies, the average bike rider can sustain a power output of 150 W for two hours and 225 W for about 30 min.

We fabricated a pedal-powered pump based on an exercise stand (Fig. 6-18). The pump develops sufficient suction to draw from shallow wells and is flexible in the sense that almost any bicycle can provide the motive power. The cost of the project was around $100, made possible by the purchase of a used bicycle exercise stand and an inexpensive vane-type pump. To keep things simple, only hand tools were used in construction.

If you decide to build your own stand, the most critical parts are the tulip-shaped lugs, hollowed out to accommodate metric and U.S. standard axle nuts (Fig. 6-19). Note that these lugs thread into engagement and lock in place. The roller assembly should pivot to make positive contact with wheels of different diameters. The flywheel is optional, but does seem to smooth the pedaling.

Ideally one would use a high-quality pump with its own shaft bearings. But such pumps are expensive, either purchased as stand-alone items or with a drive motor. We opted for a $40 electric pump from Northern Tool with the rotor cantilevered off the motor bearings. Once separated from the motor, the pump shaft had no support, which meant that it had to be assembled in dead alignment with the output shaft

Figure 6-18. *Almost any bike can be used as motive power for the pump*. Tony Shelby

Figure 6-19. *Lugs that secure the bike to the stand are the most critical parts of the assembly. Slots in the lugs appear to be for hub-transmission shifter cables, something rarely seen on modern bikes*. Tony Shelby

from the roller. The roller has a 3/8-in. shaft and the pump shaft is 10 mm.

A lathe would have made it easy to fabricate the necessary coupling. But that seemed inappropriate, if we were to build something easily replicable. Granger came to the rescue with a coupling and bushings that reduced the ID to 10 mm (Fig. 6-20). Once that problem was solved, it was simple enough to fabricate a pump bracket from scrap aluminum (Fig. 6-21).

The pump spewed out a healthy stream of water and, when stopped, held its prime (Fig. 6-22).

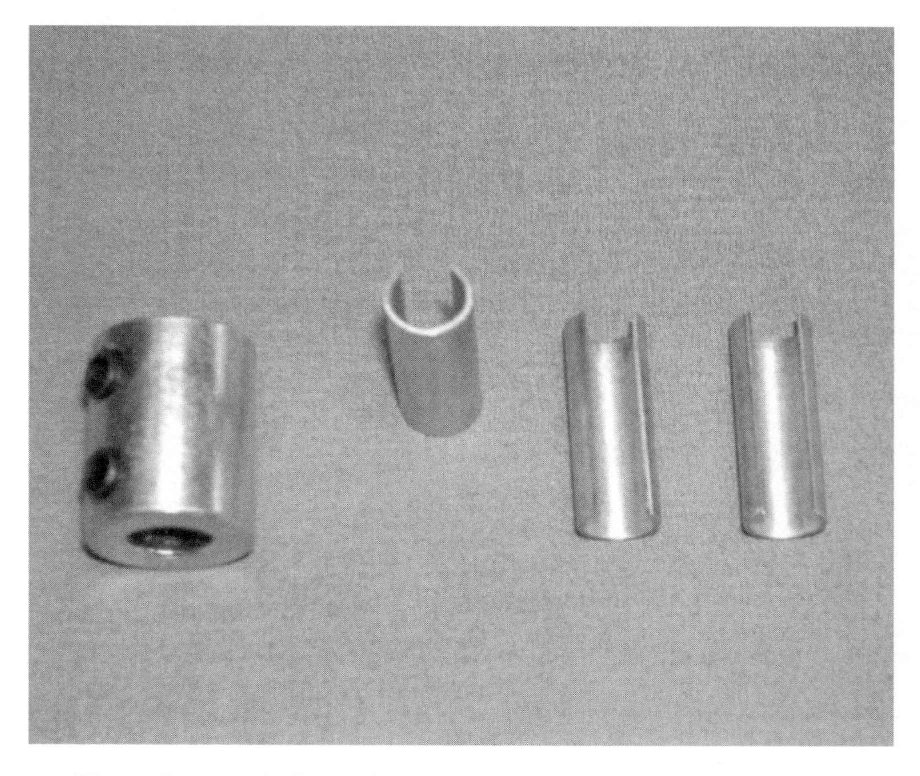

Figure 6-20. *A shaft coupling with bushings to reduce its ID.*
Tony Shelby

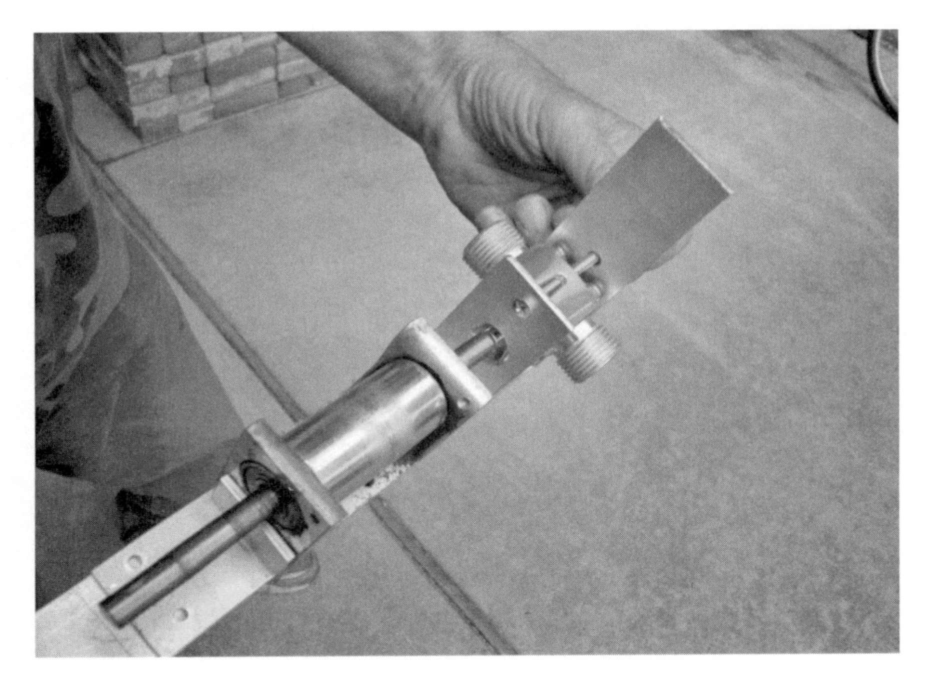

Figure 6-21. *Pump and roller assembly shown during construction.* Tony Shelby

EMAS-Style Hand Pump

What follows are instructions for building a manual pump on the general pattern of the EMAS pump, a remarkable device created 20 years ago by Wolfgang Eloy Buchner. Built from inexpensive materials, the pump lifts water 150 ft from boreholes as small as 1.25 in. EMAS is the acronym for Escuela Móvil de Agua y Saneamiento, a Bolivian school that holds classes wherever Wolfgang Buchner's duties take him.

The most critical component is the piston seal. Buchner uses the tread section of nylon-belted automobile tires. But nylon tire carcasses are difficult to find in this country, and we used a piece of conveyer belting. Almost any elastomer—the soft soles of sandals, pickup-truck bed matting, plumbing

Figure 6-22. *The pump in action with Tony providing the energy.*

washers, rubber "feet" used to dampen equipment vibration—works. Leather, the traditional seal material, may also be used.

Because of materials that were at hand, we deviated somewhat from the original design that readers with Internet access can view at http://www.youtube.com/watch?v=tUXmyv1QIS8. For example, we used light-gauge PVC, known in the trade as Schedule 20, PR 120, or PR 160 for the handle guide bushing. We also minimized the use of open flame on PVC because of concerns about toxicity.

Handle

The handle consists of a 32-in. length of ½-in. galvanized pipe threaded on both ends, two 6-in. pipe nipples, and assorted fittings (Fig. 6-23). Joints are sealed with Teflon tape, and the rough galvanized coating on the plunger are sanded smooth.

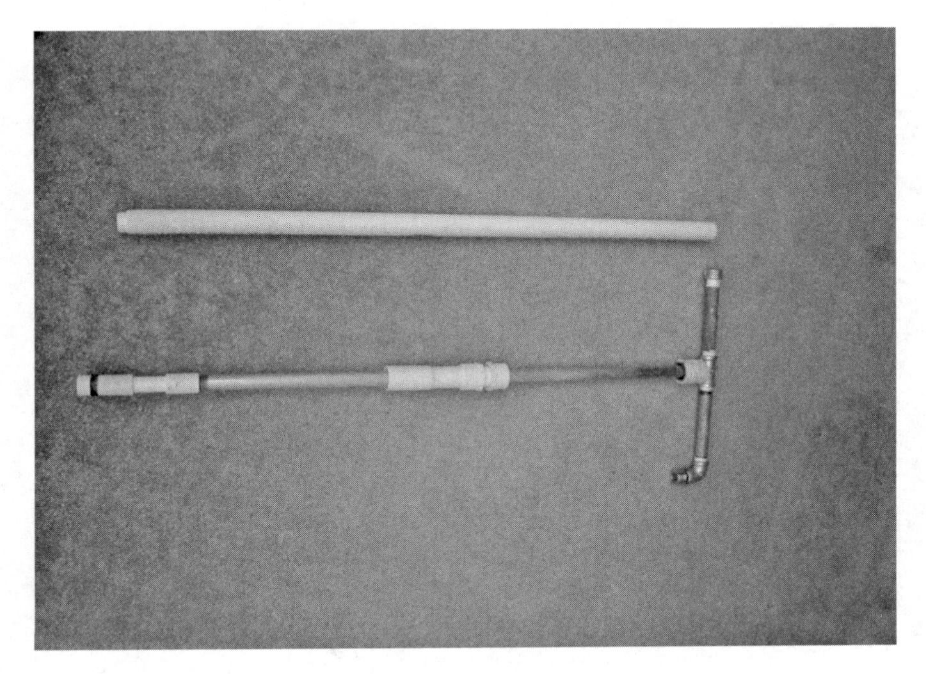

Figure 6-23. *Handle and pump barrel. The galvanized pipe plunger terminates in the piston seal and discharge check valve assembly. The guide is shown at the mid-point of the handle and functions as a bushing to keep the piston seal parallel in the barrel. The third part, just under the handle crosspiece, is an optional stop made of rubber. The lower section of the 1-in. PVC barrel is pictured below the handle. The foot valve—the inlet check valve—mounts on the bottom of the barrel. Depending upon well depth, the barrel can be as much as 150 ft long.* Tony Shelby

Check Valves

Foot and piston check valves are identical, except that the piston valve also functions as the mount for the elastomer piston seal. Each check valve consists of the following:

- One ½-in. PVC male thread × slip-fit coupling
- One ½-in. PVC female thread × slip-fit coupling; this coupling is cut down and welded to the male connector in order to provide space for the marble to move

- One glass marble
- One PVC cross piece to secure the marble.

We found three different styles of male thread × slip couplings. The best for our purposes has a squared-off shoulder at the end of the male thread. This shoulder forms a "shelf" upon which the piston seal rests (Fig. 6-24).

Construction

Step 1. File off the protrusions on both the male and female couplings (Fig. 6-25).

Step 2. Using a hacksaw, cut off the threaded end of a female coupling. Smooth and square the cut on the unthreaded part (Fig. 6-26).

Step 3. Prepare to weld the slip-fit end of the female coupling to the slip-fit end of the male coupling. We make no claims about the health risks involved in welding PVC. If you chose to do this, work outside, upwind of the fumes.

Step 4. Using a torch, heat a piece of steel or aluminum to serve as the worktable. Lightly press and twist the PVC parts on the hot metal (Fig. 6-27). Do not overheat—if the plastic smokes or turns brown, it's too hot and will not bond.

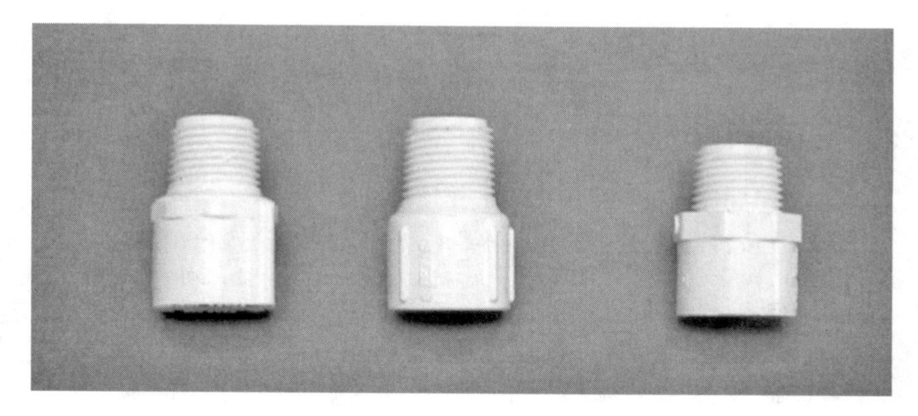

Figure 6-24. *Male thread × slip fit couplings should have a square shoulder like the one on the right.* Tony Shelby

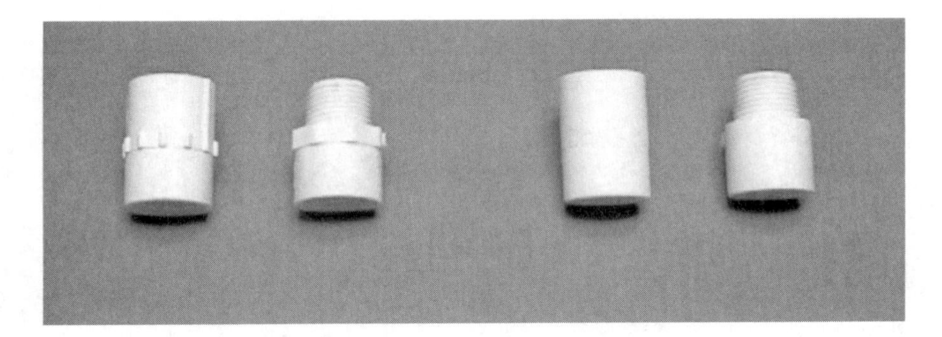

Figure 6-25. *Couplings should be filed round to move easily inside the pump barrel.* Tony Shelby

Figure 6-26. *Sand the cut smooth. The section shown on the right makes up to the male threads to retain the piston seal.* Tony Shelby

Figure 6-27. *Heating the check-valve parts in preparation for welding. Note that this operation should be performed outdoors.* Tony Shelby

Step 5. Press the parts together (Fig. 6-28). We now have a male fitting with a barrel long enough to allow the marble to rise and fall.

Step 6. Scrape and sand the PVC residue from the plate before making the next weld (Fig. 6-29). This is a safety precaution to avoid chemical boil-off and contamination of the subsequent weld.

Step 7. Cut two 3⁄16-in.-wide strips from a piece of scrap PVC pipe.

Step 8. Using a hammer, shape the end of a 20d nail to match the profile of the PVC strips.

Step 9. Insert a marble large enough to seat in the coupling. Heat the flattened end of the nail and press it

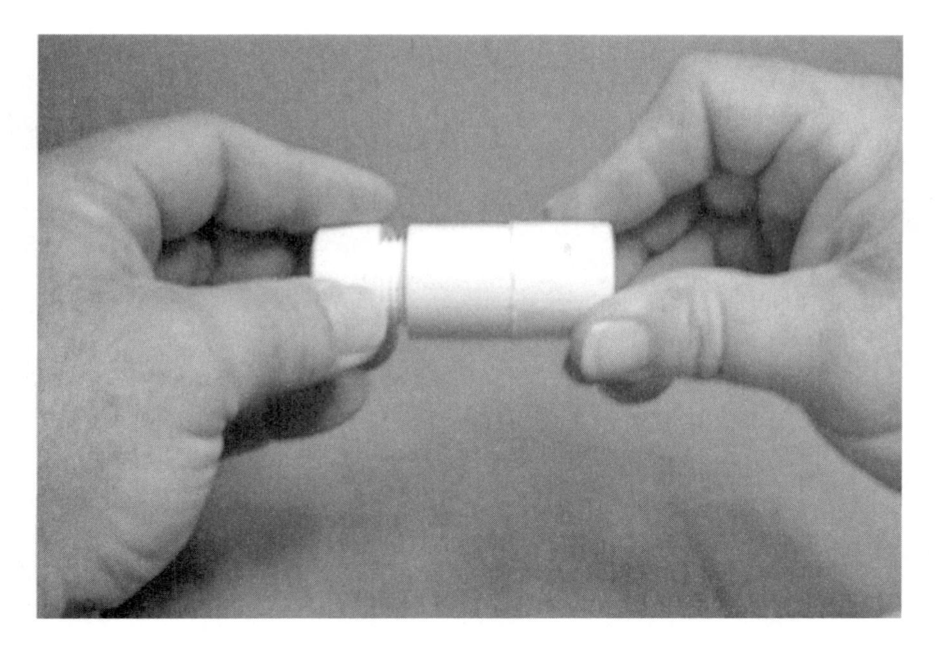

Figure 6-28. *The hot parts weld when pressed together*. Tony Shelby

Figure 6-29. *Remove the PVC residue from the melting plate before making the next weld*. Tony Shelby

through both sides of the coupling, as shown in Fig. 6-30. The holes should be just above the weld to give the marble space to move. Once the assembly cools, blow and suck on the coupling to verify that the marble seats without leakage.

Step 10. Trim the protruding ends of the pin leaving about ⅛ in. standing proud (Fig. 6-31).

Step 11. Heat the nail head and weld the pin into place (Fig. 6-32). File and sand smooth.

Step 12. Fabricate the second check valve as was done for the one already described.

Step 13. Cut the seal center hole with a ¾-in. spade bit (Fig. 6-33). As mentioned earlier, we used conveyor belting for the seal, but any ¼-in. or thicker elastomer will work.

Step 14. Using a sharp knife, rough out the seal circumference. The work goes easier if the knife is wetted. Smooth the seal edges with a sander or grinder (Fig. 6-34).

Step 15. Abrade the edges of the seal with a coarse file (Fig. 6-35).

Figure 6-30. *The flattened end of a hot nail makes rectangular holes in the PVC.* Tony Shelby

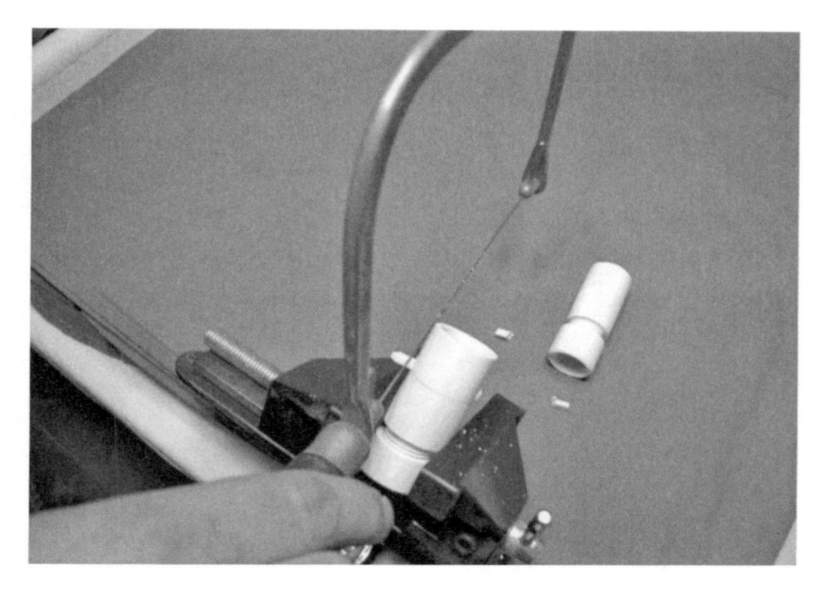

Figure 6-31. *For reasons that will become obvious in the next step, do not trim the ends of the pin flush with the coupling OD.* Tony Shelby

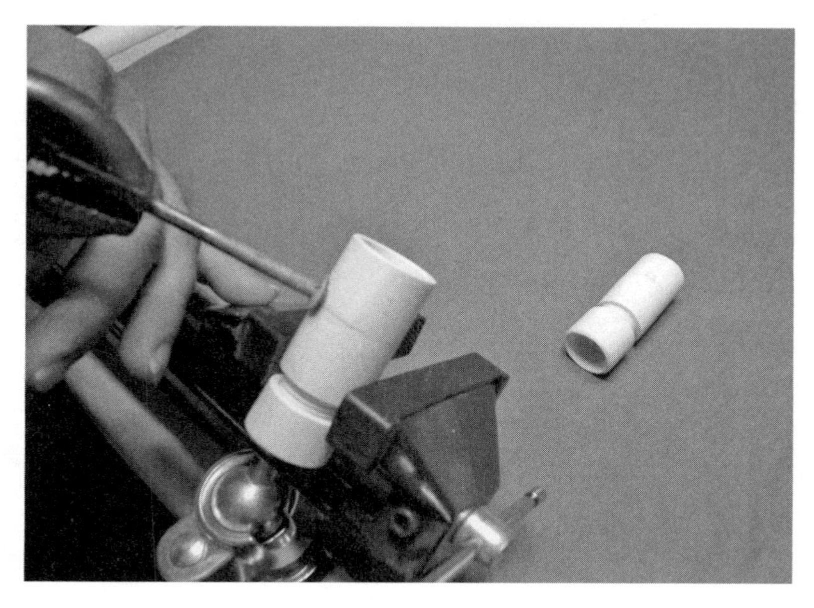

Figure 6-32. *Weld the pin into place.* Tony Shelby

Figure 6-33. *The hole in the piston seal should be sized to make a tight fit with the threaded part of the seal body. We used a ¾-in. spade drill.* Tony Shelby

Figure 6-34. *Shape the seal for a snug fit in the casing.* Tony Shelby

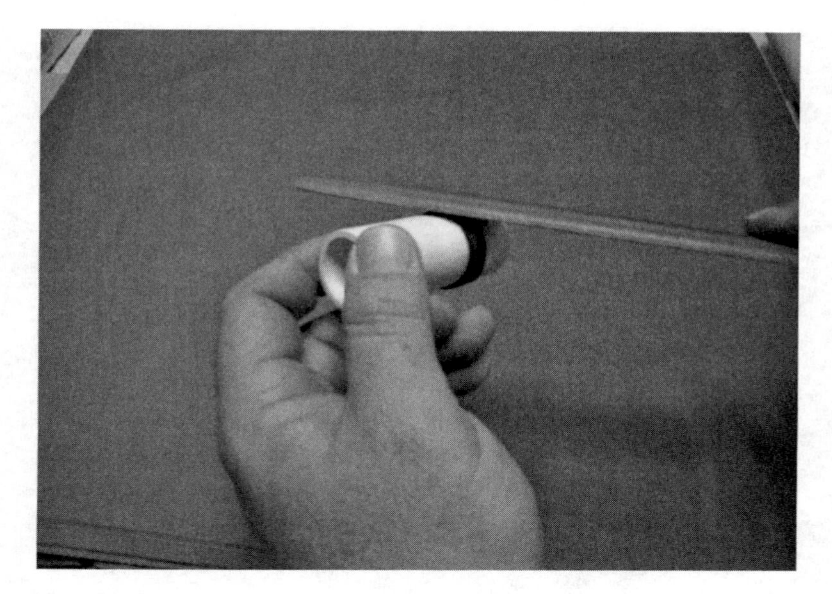

Figure 6-35. *Once the seal is shaped, roughen the edges for better contact with the barrel.* Tony Shelby

Step 16. The foot valve mounts in a heat-formed PVC guide, or bushing, that fits inside of the pump barrel. The way we fabricated it involved several operations with two grades of PVC (Fig. 6-36). The ¾-in. tapered mandrel is Schedule 40. The ¾-in. bushing and ¾-in. reducer are made of thin-wall Schedule 20 PVC known as PR 120 or PR 160. This material is used for lawn irrigation systems, air-conditioner condensate drains, and other light-duty applications. You can find it at plumbing, irrigation, and A/C supply houses and at some big-box stores.

Step 17. Bevel the Schedule 40 mandrel, as shown in the illustration above.

Step 18. Heat the ¾-in. reducer until it becomes flexible (Fig. 6-37). Insert the warm pipe over the mandrel for about the same length as the foot valve. Let the parts cool and twist off.

Figure 6-36. *From the left, a ¾-in. Schedule 40 PVC mandrel, a ¾-in. PR 120 or 160 reducer, a ¾-in PR 120 or 160 bushing, and the 1-in. Schedule 40 pump barrel. The object is to reduce the OD of the bushing enough to make a secure fit inside the barrel.* Tony Shelby

Step 19. Insert the bushing and the foot valve into the pump body (Fig. 6-38). If the parts do not fit tightly, they can be glued into place.

Note: *Both the foot valve and the piston valve install with their threads down.*

Step 20. The ½-in. galvanized pipe makes a loose fit inside of the 1-in. PVC pump barrel. There are several ways to fill in the gap. The upper end of the barrel can be heated and shrunk around the handle with a strip of rubber cut from an inner tube. We did something more complex, as shown

Figure 6-37. *Heat the bushing thoroughly and slip it into the belled end of the reducer pipe. The bell is merely a kind of funnel; we shrink the bushing diameter by pushing it past the bell and deep into the reducer. The result of this operation is a bushing with the same OD as the reducer ID.* Tony Shelby

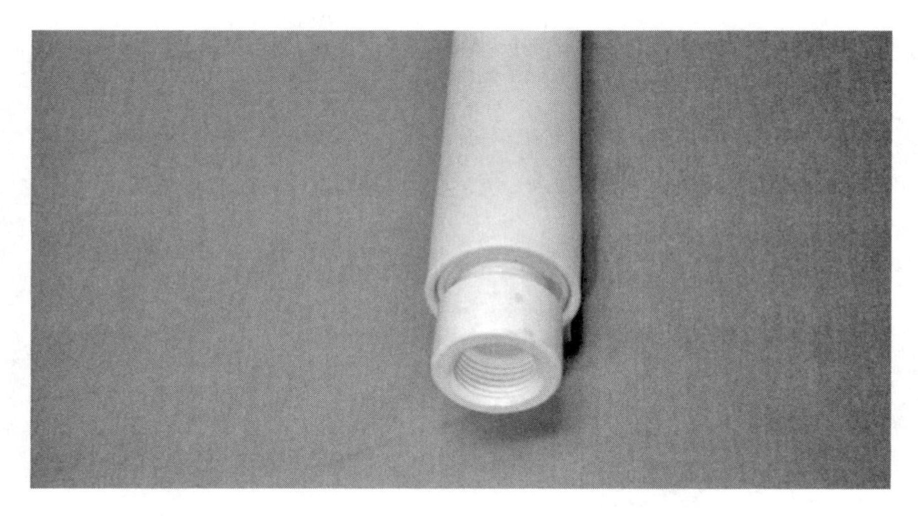

Figure 6-38. *The bushing and foot valve installed in the pump barrel.* Tony Shelby

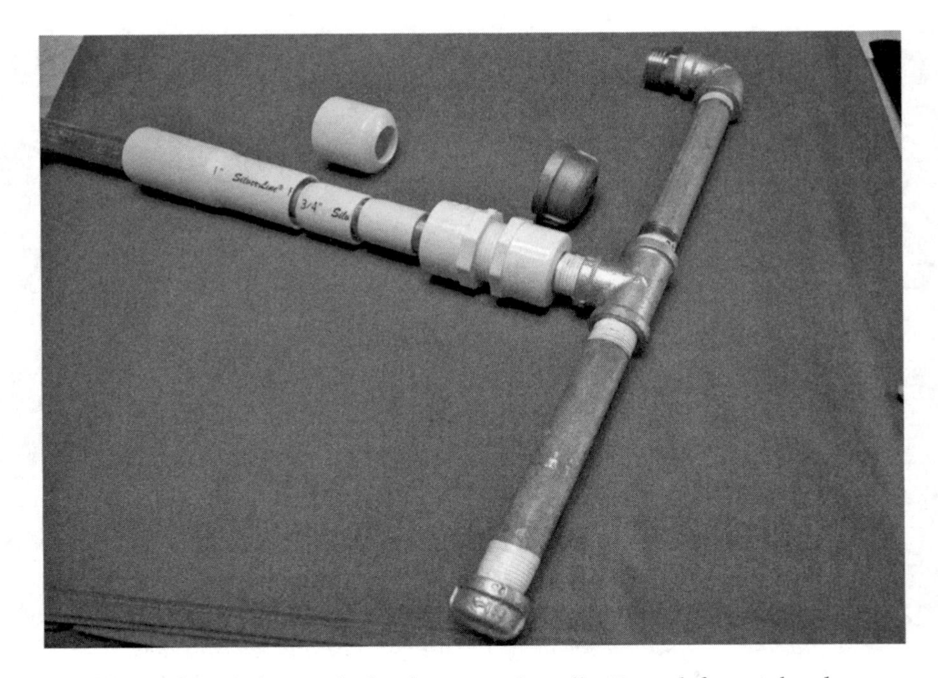

Figure 6-39. *The guide for the pump handle. From left to right, the first part is a 6-in. length of 1-in. Schedule 20 PVC. The pipe has been belled, or expanded, to fit over the 1-in. PVC pump barrel. The next is ¾-in. PR-120/160 thin-wall PVC reduced to fit inside the 1-in. pump shaft pipe. The other short piece of PVC has been reduced to fit inside the first bushing, as described in Steps 16 to 18. Initially we had planned to use a brass end cap, drilled out to handle diameter. But that was overkill. The thin plastic end cap (to the left of the brass cap) offered little by way of wear resistance. We settled on the assembly shown attached to the handle.*
Tony Shelby

in Figure 6-39. Using the same technique as for the foot valve bushing, we reduced the diameter of PR 120 tubing to serve as guide bushing for the handle.

Step 21. The pump functioned as intended (Fig. 6-40).

A Last Word

Times are changing and nothing seems off the table. The world grows warmer, grid power more unreliable, pollution

Figure 6-40. *The pump in action.* Tony Shelby

more pervasive. If this book turns out to have been worth doing, readers will roll up their sleeves and take action to assure that their loved ones will continue to have access to safe water. Harvest rainwater, drill shallow groundwater wells, investigate the possibilities of alternate power. And learn as much as you can.

While doing all this, understand that private arrangements are not enough. Water is never owned—it neither recognizes property lines nor national boundaries. What your neighbor does to the aquifer or what the feed lot down the road does to groundwater affects everyone in the area.

We must make our concerns heard. As with any DIY project, begin with the tools and skills at hand. Talk with your neighbors to find common ground. Raise your voices at city council meetings and support politicians who promise to help. Join any of the thousands of local and regional conservation organizations. Contact information for a few of these organizations—it was impossible to include more than a random sample—are listed in the back of the book.

You may also want to consider becoming a part of the Waterkeeper's Alliance. They get things done. For example, the Potomac Riverkeeper, one of nearly 200 world-wide "keeper" groups, recently was successful in forcing the city of Washington, DC, to desist from its traditional practice of discharging polluted storm water into the Potomac. The Baton Rouge-based Atchafalaya Basinkeeper put a stop to illegal logging in the endangered basin. When warned that the Basinkeeper would litigate, Carrizo Oil abandoned plans to dredge a channel in one of the last pristine areas of the Atchafalaya. This marked the first time in history that an oil company was forced to withdraw a permit for a new canal in the Basin.

The list goes on and on.

If you have any questions about this book, discover any errors in it (which may not be hard to do), or if I can be of any assistance in this battle, please contact me through the publisher.

Glossary

A (ampere or amp)—a unit of measure of the quantity of electrons flowing in a circuit.

AC—alternating current; electrical current that reverses itself at regular intervals. AC has a frequency (Hertz, Hz) of 60 cycles per second in the North America, 50 Hz in many other places in the world.

Air density—nominally 1.28 kg/m^3, or 14.7 psi, at sea level.

Anemometer—an instrument for measuring wind velocity.

Annulus—the clearance between the borehole and the casing OD. The term can also mean the clearance between the casing or borehole and the drill string.

Aquifer—an underground reservoir containing useful amounts of water.

Array (solar)—an assemblage of interconnected solar panels.

ASTM—ASTM International, formerly known as the American Society for Testing and Materials.

Autumnal equinox—the day in the fall when the sun rises and sets in 12 hours, making day and night of near equal length.

Average wind speed—the mean wind speed measured during a specified time period.

Back pumping—the technique of pumping water into a well through the drop pipe. Reversed flow clears the well screen and increases aquifer porosity.

Bentonite—a highly absorbent montmorillonite (a hydrous aluminum silicate) clay.

BLDC—brushless direct current motor; solid-state switching replaces brushes normally used in DC motors.

Bleeder valve—a valve that automatically opens to drain water from the drop pipe.

Booster pump—a pump used to increase pressure in surface plumbing.

Bowl—the pump section of a submersible well pump.

BSI—British Standards Institute.

Casing—a plastic or steel pipe installed in the wellbore to prevent collapse and shield the aquifer from contamination.

Cavitation—the formation of cavities, or bubbles, in fluids that are undergoing pumping. The bubbles violently implode leaving pits on adjacent metal surfaces.

Check valve—a valve that permits flow in one direction and blocks it in the other.

Circulation—the flow of drilling fluid from the surface to the bit and back to the surface by way of the annulus.

Cogging—the drag a permanent-magnet motor exhibits as its armature is turned.

Confined aquifer—an aquifer sealed against intrusion of surface water.

Consumptive water use—using water in ways that delay its return to surface sources.

Cp— power coefficient, or the ratio of turbine power to wind speed.

CT— composting toilet that converts bodily wastes, leaves, and other organic matter into compost.

cps—cycles per second; 1 Hz (Hertz) = 1 cps.

Cut-in speed—the wind speed at which a turbine begins to generate power.

Cut-out speed—the wind speed at which a turbine ceases to generate power.

Cuttings—sediment lifted to the surface by drilling fluid.

DC—direct, or unidirectional, current.

Diaphragm—a flexible membrane that separates the compressed air charge from the water in pressure tanks.

Discharge head—the vertical distance in feet a pump lifts water.

Discharge pipe—the horizontal pipe that conveys well water to a surface tank or outlet.

Drawdown—the distance that the static water level in a well drops during pumping, usually expressed in feet.

Drawdown capacity—the amount of water a pressure tank delivers before the pump starts. Also known as the live capacity.

Drill stem—the downhole drilling assembly, also known as the drill string.

Driller's log—the driller's work record, one copy of which is filed with state authorities.

Drilling fluid—a mixture of water and bentonite used to transport cuttings to the surface. Commonly called "mud."

Drive pipe—reinforced steel pipe used in percussion drilling.

Drive point—the point and screen assembly fitted to the end of the drive pipe; also called a wellpoint or sandpoint.

Drive shoe—a protective steel collar fitted over the lower end of casing.

Drive water—jet-pump discharge water that passes through the eductor on its way to the inlet port.

Drop pipe—the vertical pipe that carries well water to the surface, also known as the riser and less frequently as the conduit.

Eductor—the vacuum-boosting section of a jet pump; also known as the jet assembly or venturi.

FDA—U.S. Food and Drug Administration

Foot valve—a check valve mounted at or near the bottom of the drop pipe to maintain pump prime.

Friction loss—the pressure drop across piping, check valves, and other plumbing components.

gpd—gallons per day.

gpf—gallons per flush.

gph—gallons per hour.

gpm—gallons per minute.

Greywater—drainage from lavatories, washing machines, and kitchen sinks.

Grid—a centralized distribution system for delivering electrical power to consumers.

Groundwater—subsurface water.

Grout—clay or concrete filler forced under pressure between the casing and the wellbore as a seal against contamination.

GW (gigawatt)—one billion watts or 1000 megawatts.

HAWT—a horizontal axis wind turbine.

HDPE—high-density polyethylene, a semi-flexible plastic piping used for water distribution.

Head—the vertical distance that a pump can raise water; usually measured in feet.

hp (horsepower)—1 hp = 746 watts.

Hub height—the elevation of a wind turbine hub above ground level.

Humanure—composted human waste for use as fertilizer.

Hz (Hertz)—frequency expressed in seconds.

ID—inside diameter.

Impeller—the rotating element of a centrifugal pump; also called a rotor.

Insolation—the measure of the intensity of solar radiation; expressed as W/m^2.

Inverter—a solid-state device that converts DC to AC.

Jet pump—a surface-mounted pump that uses a venturi to increase its suction head.

kW (kilowatt)—a unit of energy equal to 1000 watts.

kWh (kilowatt-hour) —one kilowatt of energy generated or consumed in one hour.

LCB—a linear current booster; an electronic device that boosts current output under low voltage conditions.

Load—the power drawn from an electric circuit. Load can also refer to the power-consuming device or devices connected to the circuit.

Lost circulation—drilling fluid that does not return to the surface.

Mud— the common term for drilling fluid.

mW (megawatt)—a unit of power equal to 1000 kilowatts or one million watts.

mWh (megawatt-hour)—equal to 1000 kWh, or one million Wh.

Nacelle—the part of a VAWT corresponding to an aircraft fuselage.

NAEUS—North American End Use Study.

NEC—National Electrical Code.

NGO—non-governmental organization. As the term is used in this book, an organization that provides health services or economic assistance to developing countries.

NPSH (Net Positive Suction Head)—the maximum suction head a centrifugal pump develops before onset of cavitation.

OD—outside diameter.

Outlet—any water exit, for example, a hose bib, faucet, showerhead, or dishwasher.

Peak load—the amount of electrical power drawn during periods of maximum demand; usually expressed in terms of kilowatt-hours or megawatt-hours.

Percussion drilling—drilling with a reciprocating bit, known as driven-point drilling when done by hand and as cable-tool drilling when mechanized.

Photovoltaic cell—a semiconductor that converts light into electricity.

Pitless adapter—a subsurface cover over the drop pipe that diverts water to the horizontal discharge pipe.

Positive displacement pump—a pump that delivers the same amount of water per revolution or stroke regardless of rpm.

Potable water—water judged safe for drinking.

Power—energy converted into work over time. Commonly expressed as kWh, Btu, or hp.

Power curve—a graphical representation of the performance of a motor or pump.

Pressure, water—the force acting on water, usually expressed as psi.

Pressure switch—a diaphragm-operated switch that turns a pump on and off in response to changes in water pressure.

Pressure tank—a storage tank with a precharge of compressed air.

Prime—the presence of water at the pump inlet. Some pumps must be manually primed.

psi—pounds per square inch. 1 psi = 2.31 ft of head.

Pump controller—the switch gear for a well pump.

Pump head—the vertical distance that a pump can lift water, expressed in feet; 1 ft of head = 0.433 psi.

Pumping test—a test to determine well yield, recovery rate, and drawdown.

Pumping water level—the level of water in a well during pump operation.

PV (photovoltaic)—describes a material that is capable of generating electricity in response to light.

PV system—a collection of components for generating, storing, and delivering solar power.

Rated, or nameplate, output—the power developed by a wind turbine at its rated wind speed, commonly expressed as kW or hp.

Recharge rate—the rate at which an aquifer replenishes itself.

Rotor—the rotating element of a centrifugal pump, windmill, generator or electric motor.

Sanitary cap—a cover that fits over the wellhead.

Schrader valve—a valve that opens automatically to admit atmospheric air into a bladderless storage tank. A different type of Schrader valve permits the tank to be charged with compressed air.

SDR—Standard Dimension Ratio; the ratio of the inside diameter to the wall thickness of HDPE pipe. The lower the SDR, the thicker the pipe wall.

Short cycling—describes the rapid on/off behavior of a pump motor.

Sniffer valve—a valve that opens automatically to admit air to a bladderless pressure tank.

Solar array—multiple solar panels combined into a single assembly.

Solar cell—basic photovoltaic component grouped and interconnected to make up a solar panel, or module.

Solar module—most often a synonym for "solar panel"; the term may also mean a group of photovoltaic (PV) cells.

Solar panel—a sealed and framed assembly of interconnected PV cells.

Startup speed—the wind velocity required to put a turbine into motion.

Static water level—the vertical depth to water when the well pump is idle.

Submersible pump—a pump and motor assembly for use underwater.

Suction head—the vertical distance that a pump can draw water. Surface-mounted pumps typically develop a suction head of 25 ft or less.

Surface lift—the vertical distance from the well water level to a surface pump.

Surface water—water that collects at ground level.

Tail pipe—a lengthened intake pipe for a deep-well jet pump.

TDH—Total Dynamic Head; the vertical depth to water during pump operation, plus any additional height water must be raised, plus tank pressurization, plus frictional loses.

Tracker—a device that automatically keeps a solar array pointed at the sun.

Tremie pipe—a pipe used to pack a well with gravel and seal the annulus with grout or cement.

Turbine—a device that converts air or fluid flows into mechanical motion.

Unconsolidated formation—a formation that consists of loose, uncemented material such as sand, gravel, clay, or topsoil.

UL—Underwriters Laboratories.

V (volt)—the unit of measure of electrical potential.

VAWT—vertical-axis wind turbine.

Venturi—an obstruction to fluid flow that increases flow velocity and reduces its pressure.

Vernal equinox—the day in the spring when the sun rises and sets in 12 hours, making day and night of nearly equal length.

Vertical lift—the vertical distance a pump either can or is required to raise water.

Voltage—the measure of electromotive force, analogous to pressure in a hydraulic circuit.

W (watt)—a unit of electrical power: 1W = 1/1000 kW.

Wake losses—the loss of air velocity and flow cohesion downwind of a turbine.

Water table—the upper surface of an unconfined aquifer.

Well development—the use of reverse flow, compressed air, or other techniques to increase the permeability of the aquifer.

Wellhead—the top of a well, normally sealed with a sanitary cap.

Wind power class—available wind power on a scale of 1 to 7, with 7 being the highest.

Wind shear—in the context of wind turbines, wind shear refers to the increase in wind speed at hub height.

Xeriscape—a collection of gardening techniques that reduce water requirements.

Yield—the rate of water delivery from a well, usually expressed in gallons per minute (gpm) or gallons per day (gpd).

APPENDIX

Water Web Sites

General

http://pubs.usgs.gov/gip/gw_ruralhomeowner/gw_rural homeowner_new.html—groundwater, aquifers, and other important material.

http://www.cd3wd.com/cd3wd_40/cd3wd/index.htm—hundreds of free, downloadable manuals on water-related topics—everything our book is about.

http://www.dsireusa.org—database of government and utility alternative energy incentives.

http://www.terrylove.com/forums/—plumbing, wells, and garden irrigation.

Drought

http://www.cpc.ncep.noaa.gov/products/expert_assessment/DOD.html—United States' drought outlook, given by region and by state.

http://www.watercrunch.com/—drought monitor.

Conservation

http://water.epa.gov/infrastructure/sustain/index.cfm—EPA sustainable water website.

http://waterfortheages.org/how-you-can-make-a-difference/—saving water at home.

http://www.h2oconserve.org/home.php?pd=index— similar to above.

Composting Toilets

http://www.biolet.com/resources/documents/epa-composting-toilet-fact-sheet.pdf—general introduction to CTs.

http://www.nytimes.com/2011/09/27/health/27toilet.html—biodegradable toilet.

Contamination and Purification

http://water.usgs.gov/owq/FieldManual/index.html—a field manual for collecting water-quality data.

http://www.cdc.gov/healthywater/drinking/travel/backcountry_water_treatment.html—neutralizing protozoa, bacteria, and viruses originating from fecal matter.

http://www.drinking-water.org/html/en/Treatment/Filtration-Systems-technologies.html#tech——important source on filtration.

http://www.pollutionissues.com/Na-Ph/Nonpoint-Source-Pollution.html—non-point source pollution.

http://www.seewaldlabs.com/water-treatment-information—Penn State info on water testing, methane gas, and pollution from nearby gas-well drilling.

http://www.seewaldlabs.com/images/Literature/GasWellDrilling/XH0020.pdf—chlorine shock for contaminated well water.

http://www.sodis.ch/methode/anwendung/ausbildungs material/dokumente_material/manual_e.pdf—describes the use of plastic bottles, pathogens, limitations of process, how to identify the type of plastic, and so on.

http://www.tngun.com/pool-shock-for-water-purification/—water purification using calcium hypochlorite and information about laundry bleach.

Wells

http://manualwelldrilling.org/wp-content/uploads/2011/07/Manual_Well_Drilling_Manual.pdf—manual for drilling wells by hand.

http://ga.water.usgs.gov/edu/earthgwwells.html—wells, aquifers, and more.

http://web.archive.org/web/20091026180803/http://www.geocities.com/leonvida/Leonvida/Logros.htm—drilling by hand. This Web site is in Spanish.

http://www.drillyourownwell.com/index.htm—well drilling and jetting DIY wells; an excellent site.

http://www.endot.com/support/endopure/1—poly drop pipe.

http://www.inspectapedia.com—comprehensive source for water wells and related equipment.

http://www.michigan.gov/documents/deq/deq-wb-dwehs-wcu-gwfundwwinspmanual_221322_7.pdf—Michigan State water well manual; comprehensive.

http://www.michigan.gov/documents/deq/deq-wd-gws-wcu-wellsurvey_270629_7.pdf—charts on well depths, casing types, drilling techniques.

http://www.siminet.org/images/pdfs/watertech-latinamerica.pdf—various well-drilling, storage, and filter techniques used in Latin America.

http://www.watersystemscouncil.org/VAiWebDocs/WSCDocs/2567958WSC_INST_20.pdf—information on sizing a well pump.

Electric Pumps

http://www.greenroadfarm.com/wells.html—submersible pumps.

http://www.pumpsonline.com—excellent, no-nonsense source.

http://www.watersystemscouncil.org/VAiWebDocs/WSCDocs/2567958WSC_INST_20.pdf—sizing a well pump.

http://www1.agric.gov.ab.ca/$department/deptdocs.nsf/all/agdex719—submersible pumps.

Solar Pumps

http://www.builditsolar.com/Projects/WaterPumping/waterpumping.htm—multiple projects.

http://www.cprl.ars.usda.gov/remm-publications.php—USDA papers on solar, wind, and hybrid well pumps.

www.oregon.gov/ENERGY/RENEW/Solar/docs/pvbasics.pdf—PV basics.

www.wbdg.org/ccb/DOE/TECH/wksafe.pdf—working safely with solar.

Windmills

http://www.omafra.gov.on.ca/english/engineer/facts/03-047.htm—electricity generation using small wind turbines.

http://www.derm.qld.gov.au/factsheets/pdf/water/w44.pdf—direct-drive windmills.

http://aermotorwindmill.com—supplier of direct-drive windmill pumping systems.

http://scoraigwind.co.uk/—Hugh Piggott's blog is must reading for anyone interested in windmills.

http://windempowerment.org/forum/—DIY windmill forum.

http://windempowerment.org/wp-content/uploads/Blade-Element-Momentum-analysis-of-pvc-pipe-VS-wood-blades-on-a-100W-turbine.pdf—wind-tunnel tests of lift and drag forces on PVC and carved wooden blades.

http://www.old.windmission.dk/workshop/BasicBladeDesign/bladedesignleft.html#anchor292796—simplified blade calculations. Note: A new Web site is being built, so the site address will change to http://www.windmission.dk.

http://www.wot.utwente.nl/publications/technical_report_1990_diever_450.pdf—construction manual for a windmill with the engineering calculations.

https://www.dmr.nd.gov/ndgs/Newsletter/NL99S/PDF/windenergys99.pdf—Wind energy and a map of U.S. wind potential.

Human-Powered Pumps

http://www.recpro.org/assets/Library/Parks/omguidehandpumps.pdf—operation and maintenance manual for pitcher pumps.

http://blip.tv/mobile-school-for-water-and-sanitation/pumps-emas-standard-handpump-2465336—instructions for making an EMAS standard hand pump.

http://usfmi.weebly.com/uploads/5/3/9/2/5392099/7_-_how_to_select_the_proper_human-powered_pump_for_potable_water_ed_stewart.pdf—how to select a hand-powered pump.

www.emas-international.de—EMAS home page.

Greywater

http://www.oasisdesign.net/greywater/misinfo/index.htm—a comprehensive study of greywater.

Plumbing

http://www.buildingadream.com/techbriefs/cpvccopper.htm#Health%20CPVC—CPVC vs. copper pipe.

http://www.harvel.com/sites/www.harvel.com/files/documents/Harvel-PVC_CPVC_Pipe_Installation_Guide.pdf—PVC and CPVC installation manual.

http://www.endot.com/support/endopure/1—information on poly drop pipe.

NGOs

http://thewaterproject.org/—an NGO that focuses on African water infrastructure.

http://www.hydromissions.com/—a missionary group that offers plans, tools, and classes.

http://www.wateraid.org/—an NGO that does water projects exclusively.

http://www.waterkeeper.org/—Water Keeper Alliance, an extremely effective NGO with its main focus on U.S. waterways.

http://www.wecf.eu/english/publications/2010/annual report-2009.php—Women in Europe for a Common Future.

http://www.wot.utwente.nl/en/—a Dutch student organization with some very creative ideas.

Index

D

H

I

P

T